Agricultural Landscape Stability and Sustainable Land Management

Agricultural Landscape Stability and Sustainable Land Management

Editor

Yu Cao

MDPI • Basel • Beijing • Wuhan • Barcelona • Belgrade • Manchester • Tokyo • Cluj • Tianjin

MDPI

Editor
Yu Cao
Zhejiang University
Hangzhou, China

Editorial Office
MDPI
St. Alban-Anlage 66
4052 Basel, Switzerland

This is a reprint of articles from the Special Issue published online in the open access journal *Sustainability* (ISSN 2071-1050) (available at: https://www.mdpi.com/journal/sustainability/special_issues/Agricultural_Landscape_Land_Management).

For citation purposes, cite each article independently as indicated on the article page online and as indicated below:

LastName, A.A.; LastName, B.B.; LastName, C.C. Article Title. *Journal Name* **Year**, *Volume Number*, Page Range.

ISBN 978-3-0365-8082-1 (Hbk)
ISBN 978-3-0365-8083-8 (PDF)

Cover image courtesy of Yu Cao

Contents

About the Editor

Yu Cao

Yu Cao, Ph.D., Professor, Department of Land Management, School of Public Affairs, Zhejiang University (since 2006) and Land Academy for National Development, Zhejiang University (since 2014). He received his Ph.D. in ecology from the Institute of Applied Ecology, CAS in 2003. He carried out his postdoctoral research work in the Institute of Geographic Sciences and Natural Resources Research, CAS from 2003 to 2006. He was a visiting scholar in the Department of Geography and Geographic Information Science, University of Illinois at Urbana - Champaign from 2009 to 2010. Now, Dr. Cao serves as an editorial advisory board member of the Chinese Journal of Ecology, a council member of the International Association for Landscape Ecology China Branch and Society of China Land Science, respectively. He has also served as a vice director of the Department of Land Management, Zhejiang University, since 2011. Dr. Cao has been focusing on fundamental scientific issues and applied research on management decision making in the fields of landscape ecology, land resource management, and ecosystem management, especially within rural areas in China. This research is mainly based on the interdisciplinary integration of landscape ecology and related land resource sciences, responding to China's major national development strategies and realistic needs such as cropland protection and food security, rural revitalization, and ecological civilization. His major research interests include the multifunctional conservation of croplands and management of rural landscape sustainability, land comprehensive consolidation and ecological restoration. He has published more than 60 peer-reviewed papers in major academic journals and more than 10 scientific monographs and textbooks.

Preface to "Agricultural Landscape Stability and Sustainable Land Management"

Agricultural landscapes have multifunctionality that not only ensures human food needs but also has ecological functions. The sustainable utilization of agricultural landscapes is currently a research hotspot. Although many studies have examined the evolution of the spatial pattern of agricultural landscapes, the effects of changes in agricultural landscape functions and environmental effects on sustainable development under the influence of human activities have not been fully understood, the exploration of the optimization path of agricultural landscape spatial patterns is relatively lacking, and it is still unclear how property rights system reform can promote the effective utilization of agricultural landscapes. This has drawn lots of attention from researchers worldwide. To promote its further progress, the book focuses on the theme of agricultural landscape stability and sustainable land management, covering topics related to environmental, social, and economic dimensions. Inside, a group of researchers has made some contributions.

This book includes nine original research articles devoted to a variety of subjects including ecosystem services in the agricultural landscape, comprehensive rural development, cultivated land protection, ecological compensation for cultivated land occupation, rural spatial planning, regional governance, rural land ownership reform, and remote sensing monitoring. In the context of agricultural ecological protection, Ordoñez et al. assessed the ecosystem service values in a water supply basin, and Li et al. assessed the ecosystem services of cultivated land and constructed ecological compensation standards for cultivated land protection. Su et al. focused on the comprehensive land consolidation projects widely carried out in rural areas of China; these include constructing and optimizing ecological networks to solve problems in rural areas, such as cultivated land fragmentation, scattered spatial patterns of construction land and ecological environment pollution, and boosting the rural revitalization strategy. Wang et al. found that improving farmers' perceived benefits and reducing perceived risks is conducive to improving farmers' perceived value of cultivated land quality protection. Xia et al. found that the pastureland rehabilitation program has improved the life satisfaction of Tibetan herdsmen. Santos et al. found that technification in dairy farms may reconcile habitat conservation in a Brazilian Savanna region. Zhang et al. and Yuan et al. revealed the driving mechanism and impact of the reform of China's farmland and homestead property rights system. Yu et al. developed a high-accuracy method to extract arable land using effective data sources.

We offer our special thanks to Jiayi Wang, Xiaoqian Fang, and Guoyu Li of the Zhejiang University for their efforts in promoting the publication of papers in this Special Issue, as well as the MDPI publishing and editorial staff for their excellent work.

In future research, we will pay more attention to the measurement of sustainability, management of agricultural landscapes, evaluation methods for agricultural landscape ecosystem services, regional governance of agricultural landscapes, and so on.

We hope the readers will find this book to be informative and that the complementary research efforts will bring forth new ideas and research works on different related aspects. This should be of great significance for promoting further development in this field.

Yu Cao
Editor

sustainability

MDPI

Article

Optimizing the Compensation Standard of Cultivated Land Protection Based on Ecosystem Services in the Hangzhou Bay Area, China

Hua Li [1], Dan Su [1], Yu Cao [1], Jiayi Wang [1] and Yu Cao [1,2,*]

[1] Department of Land Management, School of Public Affairs, Zhejiang University, Hangzhou 310058, China; 22122132@zju.edu.cn (H.L.); su_dan@zju.edu.cn (D.S.); caoyu98@zju.edu.cn (Y.C.); jiayi.wang@zju.edu.cn (J.W.)

[2] Land Academy for National Development, Zhejiang University, Hangzhou 310058, China

* Correspondence: caoyu@zju.edu.cn

Abstract: The significant positive externality of cultivated land ecosystem services leads to the low comparative benefit of cultivated land utilization and then causes practical problems such as the abandonment and non-agriculturalization of the cultivated land, which poses a threat to China's food security. The existing protection system only focuses on the quantity requirement and food production service of cultivated land and ignores the multi-function of cultivated land as an ecosystem, resulting in insufficient incentives and poor effect. Therefore, it is necessary to optimize the protection's economic compensation standard by adding the cultivated land's ecosystem service value in order to comprehensively assess cultivated land resources and correct for externalities. Taking the area around Hangzhou Bay, where the contradiction between cultivated land protection and economic development is prominent, as an example, the values of six typical cultivated land ecosystem services in 2016 was constructed and calculated, including food production, carbon sequestration and oxygen production, water conservation, soil conservation, biodiversity maintenance, and cultural leisure. Combined with ecosystem services' values and the quality index, we finally determined the new county-level compensation standard of cultivated land protection in the Hangzhou Bay area. The results show that the value of cultivated land ecosystem services present obvious regional disparities, meaning that there exist significant differences in the sustainable use capacity of cultivated land and the necessity of establishing grading compensation standards in the region. Finally, we analyze the rationality and innovation of the new compensation standard model as well as its role in the protection of cultivated land and look forward to promoting the sustainable use of cultivated land through these new incentives.

Keywords: cultivated land protection; ecosystem services; ecological compensation; compensation standard; Hangzhou Bay

Citation: Li, H.; Su, D.; Cao, Y.; Wang, J.; Cao, Y. Optimizing the Compensation Standard of Cultivated Land Protection Based on Ecosystem Services in the Hangzhou Bay Area, China. *Sustainability* **2022**, *14*, 2372. https://doi.org/10.3390/su14042372

Academic Editor: António Dinis Ferreira

Received: 31 December 2021
Accepted: 16 February 2022
Published: 18 February 2022

1. Introduction

As an important carrier of food production, cultivated land is an indispensable resource for social stability, and the protection policy of cultivated land has risen to the height of China's national strategy [1]. However, compared with construction land, the comparative benefit of cultivated land utilization is relatively low due to the positive externality, which cannot get economic returns in the market [2]. As a result, the problem of cultivated land protection is considerably prominent. For example, severe cultivated land abandonment has generally occurred in economically developed areas, while at the same time, the main grain-producing areas are suffering from the degrading quality and continuous capacity overdraft of cultivated land due to unreasonable and extensive utilizations such as a high multiple cropping index, the excessive use of chemical fertilizers, pesticide pollution, etc. [3–7]. Cultivated land protection is facing the real dilemma of insufficient incentives, serious pollution, and ecological destruction [8].

As a semi-artificial and semi-natural ecosystem, cultivated land not only has the economic function of food production, but also has other important ecosystem services (ESs), such as carbon sequestration, oxygen production, water conservation, biodiversity maintenance, etc. [9–11]. These valuable tangible products or intangible effects that can be used by humans and are provided by cultivated land ecosystem are the ESs of cultivated land. The Millennium Ecosystem Assessment (MA) divides ecosystem services into four categories: supply, regulation, support, and cultural services [12]. With reference to the classification of MA, existing studies have further divided cultivated land ESs into: supply services for the provision of food, raw materials, and energy; regulation services for climate, hydrology, and disease control; support services for soil formation and habitat support; and cultural services for leisure travel and homeland complexes [13–15]. Treating cultivated land as a multifunctional ecosystem helps to comprehensively manifest its value and promote the scientific and sustainable utilization of cultivated land resources [16].

Institutionally, China's cultivated land protection system has always focused on "constrained protection" under the implementations of binding policies, such as red line delineation and the requisition–compensation balance for cultivated land [4]. However, insufficient attention has been paid to cultivated land "incentive protection" in reality [17]. This has led to the protection of cultivated land becoming a passive political task rather than an active social responsibility, due to its high cost and low benefit. In areas with large populations and rapid urbanization, it is quite possible for the local government to destroy or abandon cultivated land in pursuit of economic development. Existing research shows that the expansion processes of 60 relatively developed cities in China over the past 40 years (1973–2013) have occupied a total cultivated land area of 8426.46 km^2, accounting for 56.50% of the actual urban expansion area [18]. Moreover, research indicates that 12% of rural households have abandoned farmland in China, with an average area of 0.33 mu, accounting for 6.22% of household contracted land area [19]. That is to say, the conflict between cultivated land protection and economic development has intensified with the increasing income gap between the agricultural and non-agricultural sectors.

In the face of the severe cultivated land protection situation, however, the relevant policies are not completely effective. The current compensation policy for cultivated land protection in China, to our knowledge, has mainly focused on economic compensation under the framework of the environmental protection plan and agricultural subsidies [20], which leaves a lacuna to research the compensation mechanism considering the cultivated land ecosystem service value (ESV). The current compensation policy for cultivated land protection has the following limitations: (1) it only focuses on the food production service and overemphasizes quantitative indicators; (2) it neglects the other ESs of cultivated land, leading to a pattern of unsustainable use and a severely underestimated compensation standard, which further results in destructive behavior and reduced protection enthusiasm; (3) its current setting is relatively unfounded and outdated. The province or the whole city adopts a unified and fixed compensation standard that is not linked to the protection performance and efforts; and (4) the non-graded compensation standard makes the policy's incentive effect overly simple and short-term, and it has even gradually evolved into a formalist compensation [21–24].

Compared to foreign cultivated land protection and compensation policies, the absence of ecological elements and grading standards is an obvious problem in China. For example, the U.S. Land Conservation Reserve Program (CRP) cooperates with the Environmental Benefits Index (EBI) method to form a compensation mechanism for cultivated land's ecological protection. The evaluation indicators included in the EBI incorporate biodiversity, soil erosion, water quality, and other basic environmental elements [25]. The German cultivated land compensation standard usually refers to the nitrogen content of land and combines the landscape ecological account to achieve the balance of ESV [26]. The European Union's Environmentally Sensitive Areas (ESA) Scheme sets a national minimum level of good farming practice for each country and bases subsidies on amounts above this baseline level [27]. Canada, South Korea, and other countries have similar practices [28].

As we all know, private interest is an important driver determining subjects' behaviors according to the economic principle [29]. Thus, a reasonable compensation standard of cultivated land protection will significantly improve the enthusiasm of the related subjects to protect. The economic compensation standard is the core issue [30], which is not only related to the effect [31], but also contains the social value orientation [32]. Therefore, in the process of formulating compensation standards for cultivated land protection, adding the dimensions of cultivated land ESV can internalize the externality and increase the attention of different stakeholders of the society on the ecological protection of cultivated land [33,34], which is of great significance.

At present, there are mainly two methods to estimate ESV. One is the equivalent factor method based on Costanza's division of ecosystem function, and the study of Xie is the most relevant in China [35–37]. The other is the functional value method, which establishes an appropriate model for each service, such as the shadow engineering method, the constrained variable metric methods (CVM), the willingness to pay method (WTA), and so on [38–40]. At present, the computational research on ESV is more concentrated on forests, grasslands, watersheds, and nature reserves, and there are fewer ESV studies on cultivated land [41]. Most of the research on cultivated land ESV is focused on the necessity of compensation and the definition of compensation subjects at the theoretical level, while less research is done at the quantitative level [22]. The equivalent factor method is mostly used due to its low difficulty of data collection, but the results are highly subjective [42,43].

In order to manifest the value of cultivated land resources and emphasize ecological protection, we would like to optimize the compensation standard of cultivated land protection based on ESV. This paper takes the Hangzhou Bay (HZB) area, a typical region where there are conflicts between cultivated land protection and economic development resulting from the intense urbanization process, as a case study. We constructed a framework to analyze the ESs of cultivated land and a new model to estimate the value. Herein, we try to use the comprehensive model (combining equivalent factor method and functional value method) to further define the ESV of cultivated land, making the results more thorough and reasonable. Then, we added the cultivated land quality index for correction and finally determined the new compensation standard of cultivated land protection based on ESV. Our work is expected to provide scientific guidance for the formulation of compensation standards, improve the incentive effect of compensation policies, and promote the sustainable use of cultivated land resources.

2. Materials and Methods

2.1. Study Area

The Hangzhou Bay (HZB) area is located on the eastern coast of China and includes five prefecture-level cities in Zhejiang Province (Hangzhou, Ningbo, Jiaxing, Huzhou, and Shaoxing) (Figure 1a). The entire area is distributed along the Hangzhou Bay waters, covers an area of 44,988 km^2, and is the most economically, socially, and culturally developed and active area in Zhejiang Province—maybe even in eastern China. According to the city planning of Zhejiang Province in 2012, it is a vital strategy to accelerate the construction of the cities in the HZB and promote it to become an important part of the south wing of the world-class urban agglomeration in the Yangtze River Delta. By the end of 2020, the total population and GDP of the five cities in the HZB was approximately 35.38 million and 4322.6 billion yuan, accounting for 55% and 68%, respectively, of Zhejiang Province, while the land area is only 42% of the province. It can be seen that the HZB, with a high level of urbanization, dense population, and agglomeration of economic activities, is one of the representative areas for rapid development in China.

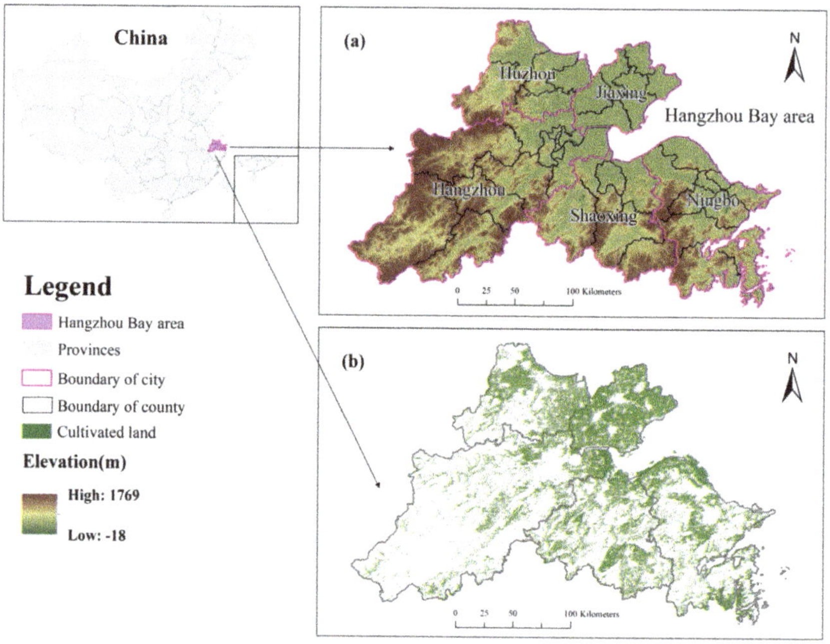

Figure 1. The geographical location (**a**) and cultivated land resources (**b**) of the study area.

At the same time, because it is located in the plains of the middle and lower reaches of the Yangtze River, there are abundant cultivated land resources (Figure 1b). The HZB area is an integrated agricultural region in history, and the Hangzhou-Jiaxing-Huzhou Plain and the Ningbo-Shaoxing Plain are famous as granaries. After entering the industrial civilization, HZB rapidly became a highly concentrated area of economic activities due to its rich resources and policy planning, leading to an over-expansion of construction land and a reduction in cultivated land, so the contradiction between economic development and cultivated land protection is prominent. Therefore, it is of great significance to develop new cultivated land protection policies in the study area.

2.2. Data Collection

The data include the multi-source data of 2016 in the five cities. The data of cultivated land patch distribution and cultivated land quality were obtained from the natural resources management department; the data of crop sown area, yield, and output value were obtained from the statistical yearbooks of the cities; the meteorological data were obtained from China's Meteorological Data Service Sharing Network (http://data.cma.cn, accessed on 9 March 2021); the soil data were obtained from the Chinese Soil Database (http://vdb3.soil.csdb.cn/, accessed on 16 March 2021); the monthly normalized difference vegetation index (NDVI) was extracted from remote-sensing images which were downloaded from the United States Geological Survey (USGS) (http://earthexplorer.usgs.gov/, accessed on 9 March 2021); and the maximum NDVI value of each unit was calculated by using the maximum value composite (MVC) method.

2.3. Classification of Cultivated Land Ecosystem Service

As mentioned above, cultivated land has various ecosystem services and multi-functions. The division of cultivated land ecosystem services is the basis for the service value evaluation. To cover MA's four categories introduced in the introduction, we selected the six types of ESs that are the most valuable and worthy of characterization in cultivated land ecosystem, including food production, carbon sequestration and oxygen production, water conservation, soil conservation, biodiversity maintenance, and cultural leisure (Table 1). Corresponding assessment models will be constructed according to the different characteristics of each ecosystem service to calculate its value. In addition, we determine the direct value of cultivated land ecosystem services that can be rewarded in the market, and the indirect value of services that cannot be recognized by the market under the influence of externalities in the current.

Table 1. The classification and interpretations of cultivated land ecosystem services.

Value Type	Classification of Service	Type of Service	Interpretation
Direct value	Supply service	Food production	Most fundamental and is the essential characteristic of cultivated land
Indirect value	Regulation service	Carbon sequestration and oxygen production	The product that participates in atmospheric regulation ecological process
		Water conservation	The utility of participating in the hydrologic regulation of ecological processes
	Support service	Soil conservation	Participating in the effects of soil regulation ecological processes
		Biodiversity maintenance	As a natural habitat for animals and plants
	Cultural service	Cultural leisure	As the main body of cultural and educational landscape aesthetics

2.4. Assessment of Cultivated Land Ecosystem Services

This study will synthetically evaluate the cultivated land ESV of the HZB region through the functional value method and the revised equivalent factor method. Since the cultivated land in the study area is mainly paddy and dry, the main crops are food crops, such as cereal (paddy rice, wheat, corn), beans and potatoes, and cash crops, such as rapeseed and vegetables. Therefore, the seven main crops mentioned above are taken as the research objects of cultivated land crops in this study, while other cash crops, such as tobacco, sugarcane and hemp, are not considered due to the small proportion of sown area.

2.4.1. Food Production

Food production service is the most basic service of the cultivated land ecosystem. As the most precious resource in the land system, the food production service of cultivated land enables human society to survive and multiply and is of great significance to human production and life. The quality of cultivated land is the sum of various natural factors and environmental factors that constitute cultivated land. It is specifically expressed in the level of productivity of cultivated land, the quality of cultivated land environmental conditions, and the quality of cultivated land products [44]. Based on this, we chose the standard grain yield corresponding to the cultivated land quality index to measure the food production capacity of cultivated land as Formulas (1) and (2).

$$X_i = \frac{C_i}{C_{max}} \times 100 \tag{1}$$

$$A = \sum_{i=1}^{n} \frac{X_i \times m_i}{M} \tag{2}$$

where A is the regional cultivated land food production service score, C_i is the standard grain yield of the i-th grade cultivated land (according to technical requirements of the third national land survey for cultivated land grading survey and evaluation in China, see in Table 2), and C_{max} is the standard grain yield of the highest grade of the cultivated land. X_i is the standardized score of the i-th grade of cultivated land, n is the number of types of regional cultivated land use grade, m_i is the area of the i-th grade of cultivated land, and M is the total area of the regional cultivated land.

Table 2. The Standard Grain Yields corresponding to the Cultivated Land Quality Grade.

Cultivated Land Quality Grade	Standard Grain Yield (kg/mu·Year)	Cultivated Land Quality Grade	Standard Grain Yield (kg/mu·Year)
1	>1400	9	>600–700
2	>1300–1400	10	>500–600
3	>1200–1300	11	>400–500
4	>1100–1200	12	>300–400
5	>1000–1100	13	>200–300
6	>900–1000	14	>100–200
7	>800–900	15	0–100
8	>700–800		

2.4.2. Carbon Sequestration and Oxygen Production

During the growth of crops in cultivated land, they fix CO_2 and release O_2 through photosynthesis, which plays a key role in regulating the carbon-oxygen balance in the atmosphere. Inverting the amount of CO_2 and O_2 released by the dry matter accumulated by crops in the process of light and action on cultivated land and then combining the cost of carbon sequestration and oxygen production to obtain the value of carbon sequestration and oxygen production services is a commonly used method in ecosystem service evaluation [45–47]. The calculation method is as follows:

$$Q = \sum_{i=1}^{n} \frac{B_i \times (1 - R_i)}{M} \quad (3)$$

$$V_1 = \varepsilon_1 \times Q \times C_1 + \varepsilon_2 \times Q \times C_2 \quad (4)$$

where Q is the dry matter quality of crops per unit area ($t/hm^2 \cdot a$); n is the number of crops; B_i is the economic output of the i-th crop ($t/hm^2 \cdot a$); R_i is the moisture content of the i-th crop; f_i is the economic coefficient of the i-th growing crop (Table 3); V_1 is the ESV of carbon fixation and oxygen production ($yuan/hm^2 \cdot a$); ε_1 is the amount of CO_2 absorbed per kilogram of dry matter, and ε_2 is the O_2 provided per kilogram of dry matter; C_1 is the value of carbon fixation in market (yuan/t); and C_2 is the value of oxygen production in market (yuan/t).

Table 3. The economic coefficient and moisture content of main crops.

Crops	Economic Coefficient-F	Moisture Content-R (%)
Paddy rice	0.45	12
Wheat	0.4	12
Corn	0.4	13
Beans	0.34	13
Tubers	0.7	70
Rapeseed	0.25	10
Vegetable	1	90

2.4.3. Water Conservation

At present, the methods for evaluating the value of cultivated land water conservation services mainly include the soil water storage capacity method, the regional water balance method, and the comprehensive water storage capacity model (including multi-dimensional considerations such as crop canopy interception, litter layer water content, and soil water storage) [48,49]. With reference to existing research results, this study uses a comprehensive water storage capacity model to evaluate the value of cultivated land ecosystems for water conservation services. In addition, since the growth cycle of crops on cultivated land is generally short and there is basically no litter, in the comprehensive water storage capacity method, only the crop canopy interception of precipitation and soil water storage capacity need to be considered [50].

$$Q_1 = \frac{\sum_{i=1}^{n} T_i \times S_i \times K_i}{\sum_{i=1}^{n} S_i} \tag{5}$$

$$Q_2 = w \times p \times h \times 100 \tag{6}$$

$$V_2 = (Q_1 + Q_2) \times C_3 \tag{7}$$

where Q_1 is the crop water interception of crops per unit area (m^3); n is the number of crops; T_i is the total precipitation in the growth cycle of the i-th crop (m); S_i is the sown area of the i-th crop (hm^2); and K_i is the Canopy precipitation interception rate of the i-th crop (Table 4). Q_2 is the soil water storage capacity per unit area (m^3); w is the soil moisture content; p is the soil bulk density (g/cm^3); h is the soil thickness, taking the average thickness of the soil profile in the study area (cm); V_2 is the ESV of water conservation (yuan/hm^2·a); and C_3 is the storage cost of artificial reservoirs (yuan/m^3).

Table 4. The precipitation interception cycle and canopy interception rate of various crops.

	Paddy Rice	Wheat	Corn	Beans	Rapeseed
Precipitation interception cycle (month)	4–10	10–5	4–9	4–10	10–5
K-canopy interception rate (%)	0.53%	1.48%	0.26%	0.90%	0.53%

2.4.4. Soil Conservation

The crops in the cultivated land ecosystem can effectively block rainfall and slow down runoff. Crops can play a good role in soil conservation by reducing the impact of water flow on the topsoil and slowing down the erosion of surface runoff. We use the opportunity cost method to evaluate the value of cultivated land in reducing topsoil loss [51]. The economic value of soil conservation services can be calculated based on the amount of soil conservation, soil thickness, and soil income per unit of cultivated land. The calculation method is as follows:

$$Ac = AP - Ar \tag{8}$$

$$V_3 = \frac{Ac \times D}{h \times p \times 10,000} \tag{9}$$

where V_3 is the ESV of soil conservation services (yuan/hm^2·a); Ac is the amount of soil conservation (t/hm^2·a); AP is the potential soil erosion amount; Ar is the actual soil erosion amount (according to Sun's research results on the soil conservation function of China's cultivated land production system [52]); D is the average income per unit area of cultivated land (yuan/hm^2·a, obtained by dividing the total output value of the 7 types of crops in each district by the total area); h is the thickness of the soil (m); and p is the bulk density of the soil.

2.4.5. Biodiversity Maintenance

The cultivated land ecosystem is a component of the natural habitats of animals and plants, so a cultivated land ecosystem has important services for maintaining biodiversity. However, the interference of human activities on the cultivated land ecosystem has resulted in the fragmentation of biological habitats and the reduction of risk resistance, thereby reducing the function of maintaining biodiversity. Therefore, we referred to the research methods of landscape ecology, selecting cultivated land patch fragmentation, NDVI index, and regional population density to construct a comprehensive evaluation index of the cultivated land biodiversity maintenance function and combined the equivalent factor method to obtain the value of this service [53].

The population density (PD) indicates the degree of interference of human activities to the cultivated land ecosystem. The greater the PD, the higher the degree of human disturbance to cultivated land.

$$\mathrm{NPD} = \frac{\mathrm{PD_{max}} - \mathrm{PD}}{\mathrm{PD_{max}} - \mathrm{PD_{min}}} \tag{10}$$

where NPD is the standardized population density of each county; PD is the population density of each county; PD_{max} is the maximum density of each county in the study area; and PD_{min} is the minimum population density of each county in the study area.

The Normalized Vegetation Index (NDVI) is an important indicator of biodiversity research. The higher the NDVI index, the better the vegetation coverage and biodiversity of cultivated land.

$$\mathrm{NNDVI} = \frac{\mathrm{NDVI} - \mathrm{NDVI_{min}}}{\mathrm{NDVI_{max}} - \mathrm{NDVI_{min}}} \tag{11}$$

where NNDVI is the standardized Normalized Vegetation Index of each county, and the other exponents have the same meaning as Formula (10).

The cultivated land fragmentation index (FI) represents the degree of fragmentation of the cultivated landscape. The higher the FI, the greater the interference intensity to biological activities.

$$\mathrm{FI} = \frac{\mathrm{NF} - 1}{\mathrm{MPS}} \tag{12}$$

$$\mathrm{NFI} = \frac{\mathrm{FI_{max}} - \mathrm{FI}}{\mathrm{FI_{max}} - \mathrm{FI_{min}}} \tag{13}$$

where NFI is the standardized cultivated land fragmentation index of each county; NF is the total number of cultivated land landscape patches in each county; and MPS is the average patch area of cultivated land in each county. The other exponents have the same meaning as Formula (10).

Then, we can obtain the index Y of the cultivated land ecosystem biodiversity maintenance services by adding all three (Formula (14)). It can be further combined with the equivalent factor method to measure the ESV of biodiversity maintenance services (Formula (15)).

$$\mathrm{Y} = \mathrm{NPD} + \mathrm{NNDVI} + \mathrm{NFI} \tag{14}$$

$$V_4 = \frac{1}{7} \times \lambda_1 \times \frac{Y_j}{\mathrm{Ymid}} \times \sum_{k=1}^{n} \frac{m_k \times p_k \times q_k}{M} \tag{15}$$

where V_4 is the ESV of biodiversity maintenance services per unit area of cultivated land (yuan/hm^2·a); k is the type of crop; n is the number of types of crops; p_k is the average price of the k-th crop (yuan/hm^2); q_k is the yield of the k-th crop (t/hm^2); m_k is the total area of the k-th crop (hm^2); M is the total area of n types of crops in the region; 1/7 refers to the economic value of ecosystem services for 1/7 of the value of production services provided by the unit area; and λ_1 is the equivalent factor of the maintenance services according to Xie's research [37].

2.4.6. Cultural Leisure

The cultural and leisure services of the cultivated land ecosystem refer to the cultural education, landscape aesthetics, and leisure travel services that people obtain from the cultivated land landscape. Generally speaking, the ESVs of cultural and leisure services are affected by both the supply and demand sides. The cultural and leisure services of cultivated land are more attractive to urban residents with dense populations and higher affordability [54]. We use the population density index of each region to modify the equivalent factor coefficient to evaluate the cultural and leisure service value of cultivated land.

$$V_5 = \frac{1}{7} \times \lambda_2 \times \eta \times \sum_{k=1}^{n} \frac{m_k \times p_k \times q_k}{M} \quad (16)$$

where V_5 is the ESV of cultural services per unit area of cultivated land (yuan/hm^2·a); λ_2 is the equivalent factor of cultural service; and η is the population correction coefficient, calculated by dividing the population density of each county by the median population density of all counties. The other exponents have the same meaning as Formula (15).

2.5. Calculation of Compensation Standards for Cultivated Land Protection

The food production service of cultivated land can obtain tangible products and market value; however, the intangible ecological utility of the cultivated land ecological system cannot be effectively manifested in the market. In other words, the positive externalities of ESs have caused some entities to share these services for free without paying, which has led to an underestimation of the value of cultivated land resources, and the protection of cultivated land has become a thorny event with large responsibilities and small benefits.

Therefore, in the process of calculating the compensation standard for cultivated land protection, the sum of the indirect value (including carbon sequestration and oxygen production, water conservation, soil conservation, biodiversity maintenance, and cultural leisure services value) of cultivated land should be taken as the ideal standard for cultivated land protection compensation. Furthermore, in combination with the actual situation, the food production capacity index based on the quality can be used for correction to finally calculate the result of the economic standard. Thus, we calculated the compensation standard using the following equation:

$$P = \frac{A}{A_{mid}} \times (V_1 + V_2 + V_3 + V_4 + V_5) \quad (17)$$

where P is the compensation standard per unit area of cultivated land (yuan/hm^2·a); A_{mid} is the median of food production service scores of all subjects; and the other variables came from the formulas mentioned above.

3. Results

3.1. Evaluation of Ecosystem Services

The evaluation results and spatial distribution of the six types of cultivated land ecosystem services are shown in Figure 2 and Table 5. The food production service with direct value are expressed by the dimensionless cultivated land quality index (exponent A), and the remaining five types are expressed by the value of unit area after statistics (yuan/hm^2·a). It can be seen that there are different spatial distribution patterns among various services. For example, taking food production service as a reference, the food production service of cultivated land in the eastern and northern regions of the study area is relatively high, while low in the western and southern regions. The overall trend is higher in the northeast and lower in the southwest and southeast. The carbon sequestration and oxygen production service and the cultural leisure service reflect the same distribution characteristics. On the contrary, the biodiversity maintenance service presents completely opposite characteristics. In addition, the water conservation and soil conservation services present a form that is high in the middle and low on the wings.

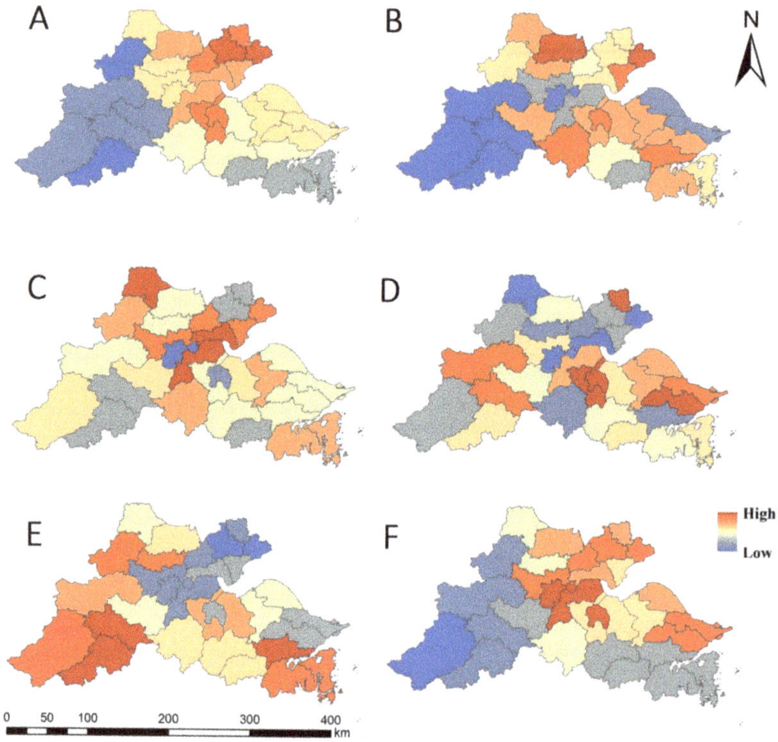

Figure 2. The spatial distribution of six ESs regarding (**A**) food production, (**B**) carbon sequestration and oxygen production, (**C**) water conservation, (**D**) soil conservation, (**E**) biodiversity maintenance, and (**F**) cultural leisure.

These gaps of cultivated land ESs between different counties are closely related to factors such as types of crops (from the calculation Formula (3), we can see that cereals, corn, and soybeans have less water content and lower economic coefficients. Therefore, they have a higher value of carbon fixation and oxygen production services compared to economic crops such as vegetables and fruits), soil thickness (directly affects water conservation and soil conservation services), cultivated land use pattern (fragmentation affects biodiversity maintenance services), and so on. Topographic factors have the greatest influence on the results, as it can be seen that the distribution of low-value areas and mountainous areas is consistent for the most part. This is because the topography affects both soil and hydrology, the key factors that make up the quality of cultivated land.

From a numerical point of view, among the five types of indirect values, the ESV of carbon fixation and oxygen production is the highest, and the gap between regions is relatively large, ranging from 1265.66 yuan/hm^2·a to 10,622.41 yuan/hm^2·a. The service value of soil conservation is relatively low in the whole region, and the value ranges from 98.96 yuan/hm^2·a to 378.69yuan/hm^2·a. The value of cultural leisure services is the lowest, which is only ranging from 7.23 yuan/hm^2·a to 137.20 yuan/hm^2·a. This reflects the degree of acceptance and the equivalent utility of different services value in the market.

Table 5. The ESV and compensation standard of cultivated land in HZB in 2016.

City	County	A *	V_1 *	V_2	V_3	V_4	V_5	P *
Hangzhou	Hangzhou District	59.28	1622.18	1458.49	100.83	743.77	137.20	4144.95
	Xiaoshan	66.37	6819.09	3547.65	233.66	708.41	123.66	13,059.75
	Tonglu	38.65	1404.84	2456.84	251.43	1407.60	21.77	3687.45
	Chunan	39.13	1801.32	2830.19	137.24	1350.93	7.23	4126.80
	Jiande	33.75	1895.22	2459.91	205.80	1379.90	17.29	3461.40
	Fuyang	39.63	8252.60	2783.85	188.70	1034.86	37.23	8388.15
	Linan	36.85	1265.66	2573.02	246.26	1258.29	17.20	3400.05
	Yuhang	60.66	6525.33	3111.29	216.19	673.17	101.71	11,096.25
Huzhou	Huzhou District	68.62	9698.83	2703.74	185.84	1192.65	64.81	16,362.83
	Deqing	62.55	8467.90	2616.56	129.66	1353.52	42.81	13,577.40
	Changxing	61.45	8583.56	3893.56	98.96	1065.37	40.43	14,470.95
	Anji	34.33	7890.53	2914.58	149.57	1297.18	22.59	7252.05
Jiaxing	Jiaxing District	85.76	8024.28	2543.84	138.72	456.48	84.10	16,628.78
	Jiashan	73.54	8035.00	2350.36	346.77	643.49	70.24	14,489.25
	Haiyan	65.91	9325.53	3160.47	143.95	944.01	59.45	15,466.95
	Haining	67.84	8043.29	3608.53	110.22	867.75	72.26	14,833.05
	Pinghu	80.30	10133.69	3396.17	99.48	425.77	81.49	19,540.05
	Tongxiang	74.21	7559.11	3132.76	134.46	664.48	87.07	14,789.40
Ningbo	Ningbo District	58.00	5204.40	2620.84	259.83	985.39	80.90	9103.05
	Yinzhou	58.10	8454.29	2724.87	315.91	909.19	99.34	12,503.55
	Xiangshan	44.67	7900.75	2926.88	213.25	1342.98	36.43	9549.30
	Ninghai	43.29	8410.60	3019.66	187.74	1306.21	31.27	9653.40
	Yuyao	57.52	8523.95	2920.42	240.69	1031.92	51.05	12,641.55
	Cixi	56.35	5228.99	2661.39	232.01	1065.11	70.52	8978.85
	Fenghua	47.52	9033.76	2675.97	127.25	1398.68	34.90	10,855.35
Shaoxing	Shaoxing District	73.03	9427.65	2019.63	304.56	975.12	137.20	16,171.35
	Shangyu	51.42	8557.91	2849.99	211.52	1242.55	50.76	11,427.75
	Xinchang	41.98	6471.30	2401.51	219.32	1154.37	32.83	7427.10
	Zhuji	50.93	9428.76	2916.59	123.31	1146.44	42.83	11,972.25
	Shengzhou	49.70	7266.74	2717.97	193.90	1199.85	37.32	9764.85
	Keqiao	71.05	8405.19	2624.06	289.77	1246.56	56.74	15,436.80

* A is the dimensionless index of cultivated land quality. V_1 through V_5 represent the five ESVs of cultivated land that we calculated (yuan/hm^2·a). P is the final compensation standard for cultivated land protection (yuan/hm^2·a).

3.2. Compensation Standard for Cultivated Land Protection

By Formula (17), the economic compensation standard for cultivated land protection in the HZB is shown in Figure 3 and Table 5 (In order to better present the relationship between compensation standard, ESV, and cultivated land quality, we also present the sum result of ESV in Figure 3). According to the research results, the average standard is 11,298.9 yuan/hm^2·a. Through the natural breakpoint method, we can realize the grading of compensation standard.

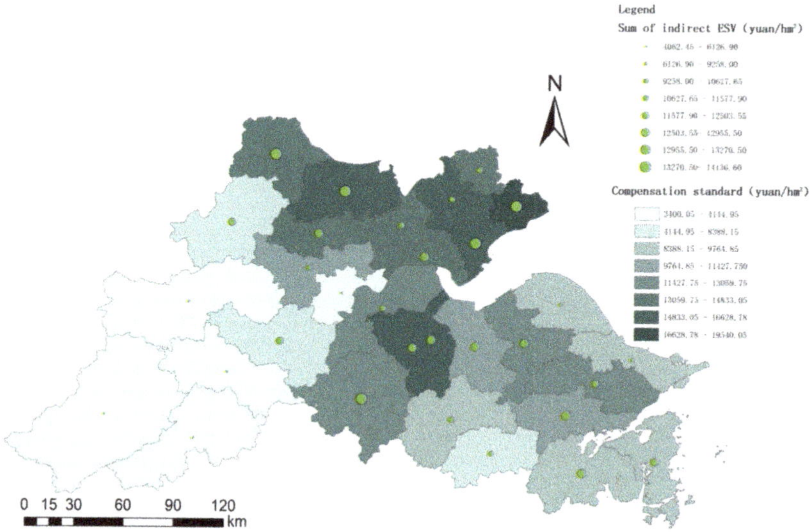

Figure 3. The spatial distribution of the sum of Indirect ESV and compensation standard.

Spatially, the low-value areas of the compensation standard are mainly distributed in city's district areas and the southwest wing of the study area, such Hangzhou district, Ningbo district, Chun'an County, Jiande County, Tonglu County, Anji County, etc. The quality of cultivated land in the municipal area is high, but the indirect ESV is low, mainly due to the serious abandonment of cultivated land; the counties in the southwest are affected by factors such as the fragmentation of cultivated land caused by topography, which leads to the low quality of cultivated land, reduction of indirect ESV, and a lower compensation standard. The high-value areas of compensation standards are concentrated in the northeastern part of the study area, which is close to Taihu Lake, where the quality of cultivated land and the utilization of cultivated land are relatively optimistic. The compensation standards of counties near the waters of Hangzhou Bay are also relatively high, with flat terrain and water networks that can enable the cultivated land ecosystem to provide better services.

Numerically, the compensation standard varies greatly among different counties. The highest compensation is more than 5 times the lowest compensation. This strongly criticizes the current "one size fits all" approach to compensation standards for cultivated land protection. By the calculation method based on ESV, we have come up with a significant grading compensation standard in HZB.

4. Discussion

4.1. The Rationality and Reality of Compensation Standard

Cultivated land has irreplaceable ecological functions; however, because of externality, the indirect value is not paid in the market. By combining the principles of ecology and economics, we added the ESV to the new calculation method of the compensation standard to partially address the market failure. In our model, the indirect value of the cultivated land ecosystem is used as the original ideal compensation standard to compensate the provision of such non-marketable environmental services and the efforts of farmers and local governments. Then, considering the dominance of the food production service in cultivated land ecosystem in reality, we chose the food production capacity based on the quality of the cultivated land as the correction index. There are three reasons why we selected this index: Firstly, the cultivated land quality index includes basic conditions for cultivation such as water, soil, fertility, and so on, and is a comprehensive indicator for judging the situation of cultivated land. Secondly, according to the quality of cultivated land, the Chinese government divides cultivated land into three categories: general farmland, basic farmland, and permanent basic farmland, and carries out different management measures according to this. So, the index we used can cater to the current value orientation of the government, giving the calculated results both ideal and realistic meanings. Thirdly, it can focus on the protection of the quality and ecological environment of cultivated land, not just on the quantity, and at the same time, achieve hierarchical protection.

Moreover, compared with previous research and current policy, the results of the compensation standard are reasonable and realistic. In the existing research, because the equivalent factor method makes it less difficult to obtain data, most studies use this method to obtain cultivated land ESV. The selection of different correction factors and various attention angles make the results highly subjective. Therefore, it is difficult to have a commonality for discussion in the studies of different scholars. Generally speaking, the calculation results of ESV for cultivated land range from a few thousand yuan to tens of thousands of yuan per hectare [42,43]. As for existing cultivated land compensation practices, the current economic compensation standards in some typical areas in China are shown in the Figure 4. Chengdu has established a cultivated land protection fund to provide 6000 yuan/hm^2 subsidies for the basic cultivated land. Suzhou's policy stipulates that the standard of cultivated land ecological compensation is 6300 yuan/hm^2. The standard of financial subsidy for basic cultivated land in Foshan City, Guangdong Province is 12,000 yuan/hm^2 and that of the Minhang District in Shanghai and the Haidian District in Beijing is 22,500 yuan/hm^2. Compared with the current standards, the range of the compensation standards for cultivated land protection calculated in this study are 3400.05 yuan/hm^2–19,540.05 yuan/hm^2. To sum up, the upper and lower limits of the compensation standard we constructed are within a reasonable range that is acceptable to the society, and the results of the study have a certain degree of reality and feasibility.

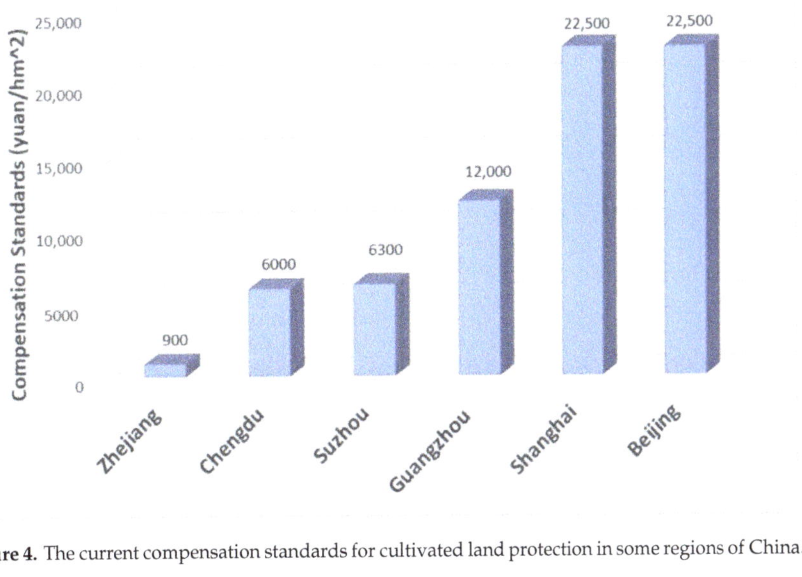

Figure 4. The current compensation standards for cultivated land protection in some regions of China.

Also, in Figure 4, it can be seen that the current compensation standard for cultivated land protection in Zhejiang Province, where HZB is located, is that the provincial finance subsidizes 450 yuan/hm^2 per year, and the municipal or the county-level financial support is not less than 450 yuan/hm^2 per year (that is 900 yuan/hm^2 per year in total), which is significantly lower than other areas. On the other hand, the policy only requires the minimum compensation standard, and the payment is based just on the area factor. This kind of low compensation standard and non-grading incentive system easily leads to a lack of enthusiasm for protection and the neglect of ecological protection. In this paper, the grading compensation standard based on ESV and the quality of cultivated land that we determined can attach great importance to the protection of cultivated land ecology and effectively realize the differentiated incentive. Under the principle of "Efforts will be rewarded," local governments, farmers, and other entities will increase their enthusiasm for cultivated land protection as it will proceed from economic driving forces and achieve better protection effects.

4.2. Insights for Cultivated Land Protection Policy

In addition to a market failure caused by "externalities," the reasons for the dilemma of cultivated land protection also include "internalities" caused by different development goals among different governments. In order to maintain social stability, the central government attaches great importance to the protection of cultivated land and has introduced severe measures to restrain destructive behavior; however, local governments, as the executors of the policies, focus more on rapid economic development. Thus, such a "principal-agent" dilemma will be resolved only when the economic compensation benefits of cultivated land protection can offset its opportunity cost. After quantifying the ecological value of cultivated land, part of the opportunity cost can be compensated so as to improve the enthusiasm of local governments.

Therefore, in the cultivated land protection system, the rights and responsibilities of each subject are divided as follows (Figure 5). The central government should act as fundraisers and regulators. It is possible to establish a compensation fund for the protection of cultivated land in the whole region through payment methods, such as the transfer of development rights (TDR) [55], and then dynamically monitoring the quality of cultivated land and regularly assessing the environmental safety of cultivated land ecosystems. Finally, the compensation funds should be allocated to local governments in a gradient according to the protection effect.

As for the local government (especially county-level governments), it should do a good job in the role of implementer and communicator. The main tasks may include managing the valuable resource of cultivated land with an ecosystem view, actively promoting the sustainable use of cultivated land in the jurisdiction through cultivated land quality improvement construction projects and penalties for illegal activities, and concurrently informing farmers of the corresponding protection policies and providing subsidy funds to them.

In addition, farmers should play an important role as the implementors of cultivated land use, and in the daily use practices of cultivated land, they should pay attention to the unified management of water sources, soil, organisms, and environment in the cultivated land ecosystem and provide practical feedback and suggestions to government departments. Only in this way can we change the protection of cultivated land from "passive" to "active".

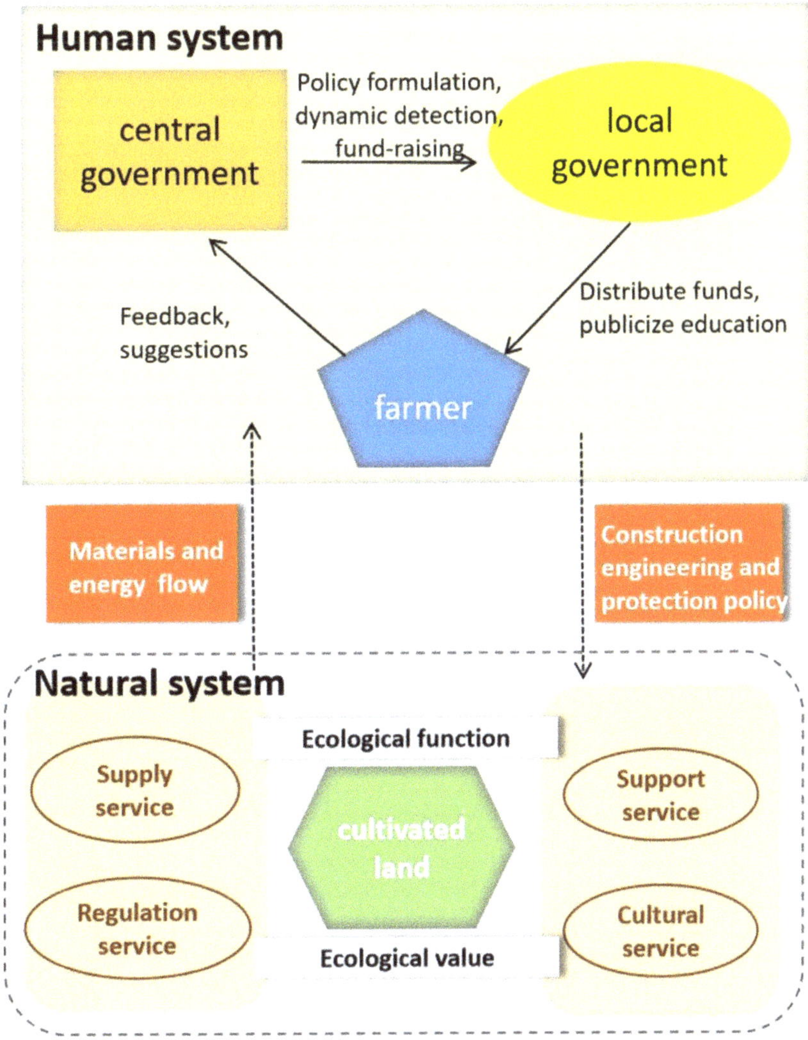

Figure 5. A conceptual map linking cultivated land ecosystems and human conservation subjects.

4.3. Limitation and Further Improvement

Most of the existing studies only use the single equivalent factor method or the functional value method to evaluate the cultivated land ESV. Here, we construct a more comprehensive method to make the results more accurate and objective. At the same time, the calculated ESV is corrected by the quality index to obtain a new compensation standard for cultivated land protection, which is of innovative significance. However, due to the availability of data, there are directions for further improvement in the research: (1) Although the cultivated land ESs this paper selected have covered the classification of MA, a more scientific and reasonable classification of cultivated land ESs is worthy of further exploration. (2) The negative value of the cultivated land ecosystem (such as air pollution, water pollution, soil pollution, etc.) has not been taken into consideration in this research. (3) The selection of the correction coefficient needs to be supplemented and improved. Especially if the research involves cross-time dimensions, correction coefficients such as the price index and the degree of socio-economic development need to be added. (4) The social value of cultivated land resources, including the guarantee of farmers' livelihood and the maintenance of social stability, is also a research focus of cultivated land protection, which is worth further study [56–58].

5. Conclusions

Using multiple data such as landscape distribution, cultivated land quality, soil depth, precipitation, and remote sensing images, we calculated the ESV of six major cultivated land ESs in the HZB region in 2016 and then used the quality index as a correction factor to obtain the compensation standard results. In summary, we pointed out the shortcomings of the existing compensation policies for cultivated land protection in terms of standard setting and value orientation and proposed a new compensation standard calculation method based on cultivated land ESV, emphasizing the importance of cultivated land ecological protection and the incentive effect of the grading compensation system.

From the research results based on the HZB case, there are obvious gaps in the ESV of cultivated land in various counties, which are closely related to the types of crops, soil and water conditions, and topographic factors, and they also reflect the differentiated contributions of these counties in the protection of cultivated land. Therefore, the compensation standard should be determined based on these differentiated contributions, so as to be fairer and more motivating. So, the optimization of the compensation standard for cultivated land protection based on ESV is an important way to combine "high incentives" with "strong constraints" in the policy design.

In the end, we suggested that the government departments should establish a comprehensive evaluation and management system of cultivated land from the perspective of ecosystem conservation. Also, the division of rights and responsibility of each subject in the cultivated land protection and compensation policy is important. While the research tries to coordinate the contradiction between cultivated land protection and economic development for the area around Hangzhou Bay, it can also provide reference for policymakers in other regions to promote the sustainable use of cultivated land.

Author Contributions: Conceptualization, H.L., Y.C. (Yu Cao, caoyu@zju.edu.cn), and J.W.; data analysis, H.L.; investigation, H.L. and Y.C. (Yu Cao, caoyu98@zju.edu.cn); methodology, H.L. and D.S.; writing—original draft, H.L. and D.S.; funding acquisition, project administration, supervision, writing—review & editing, Y.C. (Yu Cao, caoyu@zju.edu.cn). All authors have read and agreed to the published version of the manuscript.

Funding: This research was financially supported by the National Social Science Fund of China (No. 20AGL025) and the Natural Science Foundation of Zhejiang Province (No. LY19D010012).

Institutional Review Board Statement: Not applicable.

Informed Consent Statement: Not applicable.

Data Availability Statement: The data used to support the findings of this study are available from the corresponding author upon request.

Acknowledgments: Not applicable.

Conflicts of Interest: The authors declare no conflict of interest.

References

1. Ma, A.; Zhang, J. The Use of Choice Experiments to Value Public Preferences for Cultivated Land Protection in China. *J. Resour. Ecol.* **2014**, *5*, 263–271.
2. Lu, X.; Qu, Y.; Sun, P.; Yu, W.; Peng, W. Green Transition of Cultivated Land Use in the Yellow River Basin: A Perspective of Green Utilization Efficiency Evaluation. *Land* **2020**, *9*, 475. [CrossRef]
3. Su, M.; Guo, R.; Hong, W. Institutional transition and implementation path for cultivated land protection in highly urbanized regions: A case study of Shenzhen, China. *Land Use Policy* **2019**, *81*, 493–501. [CrossRef]
4. Xie, H.; Zhang, Y.; Choi, Y. Measuring the Cultivated Land Use Efficiency of the Main Grain-Producing Areas in China under the Constraints of Carbon Emissions and Agricultural Nonpoint Source Pollution. *Sustainability* **2018**, *10*, 1932. [CrossRef]

5. Zhao, J.; Luo, Q.; Deng, H.; Yan, Y. Opportunities and challenges of sustainable agricultural development in China. *Philos. Trans. R. Soc. Lond. Ser. B Biol. Sci.* **2008**, *363*, 893–904. [CrossRef]
6. Sun, M.; Huo, Z.; Zheng, Y.; Dai, X.; Feng, S.; Mao, X. Quantifying long-term responses of crop yield and nitrate leaching in an intensive farmland using agro-eco-environmental model. *Sci. Total Environ.* **2018**, *613–614*, 1003–1012. [CrossRef]
7. Hua, X.; Yan, J.; Li, H.; He, W.; Li, X. Wildlife damage and cultivated land abandonment: Findings from the mountainous areas of Chongqing, China. *Crop. Prot.* **2016**, *84*, 141–149. [CrossRef]
8. Wu, Y.; Shan, L.; Guo, Z.; Peng, Y. Cultivated land protection policies in China facing 2030: Dynamic balance system versus basic farmland zoning. *Habitat Int.* **2017**, *69*, 126–138. [CrossRef]
9. Palm, C.; Blanco-Canqui, H.; DeClerck, F.; Gatere, L.; Grace, P. Conservation agriculture and ecosystem services: An overview. *Agric. Ecosyst. Environ.* **2014**, *187*, 87–105. [CrossRef]
10. Fagerholm, N.; Torralba, M.; Moreno, G.; Girardello, M.; Herzog, F.; Aviron, S.; Burgess, P.; Crous-Duran, J.; Ferreiro-Domínguez, N.; Graves, A.; et al. Cross-site analysis of perceived ecosystem service benefits in multifunctional landscapes. *Glob. Environ. Chang.* **2019**, *56*, 134–147. [CrossRef]
11. Doxa, A.; Paracchini, M.L.; Pointereau, P.; Devictor, V.; Jiguet, F. Preventing biotic homogenization of farmland bird communities: The role of High Nature Value farmland. *Agric. Ecosyst. Environ.* **2012**, *148*, 83–88. [CrossRef]
12. Carpenter, S.R.; DeFries, R.; Dietz, T.; Mooney, H.A.; Polasky, S.; Reid, W.V.; Scholes, R.J. Millennium Ecosystem Assessment: Research needs. *Science* **2006**, *314*, 257–258. [CrossRef]
13. Zhang, W.; Ricketts, T.H.; Kremen, C.; Carney, K.; Swinton, S.M. Ecosystem services and dis-services to agriculture. *Ecol. Econ.* **2007**, *64*, 253–260. [CrossRef]
14. Cao, Y.; Cao, Y.; Li, G.; Tian, Y.; Fang, X.; Li, Y.; Tan, Y. Linking ecosystem services trade-offs, bundles and hotspot identification with cropland management in the coastal Hangzhou Bay area of China. *Land Use Policy* **2020**, *97*, 104689. [CrossRef]
15. Ou, M.; Wang, K.; Guo, J. Research progress on ecological compensation mechanism of farmland protection. *Res. Agric. Mod.* **2019**, *40*, 357–365.
16. Yu, D.; Wang, D.; Li, W.; Liu, S.; Zhu, Y.; Wu, W.; Zhou, Y. Decreased Landscape Ecological Security of Peri-Urban Cultivated Land Following Rapid Urbanization: An Impediment to Sustainable Agriculture. *Sustainability* **2018**, *10*, 394. [CrossRef]
17. Wang, K.; Ou, M.; Wolde, Z. Regional Differences in Ecological Compensation for Cultivated Land Protection: An Analysis of Chengdu, Sichuan Province, China. *Int. J. Environ. Res. Public Health* **2020**, *17*, 8242. [CrossRef]
18. Zhang, Z.; Wen, Q.; Liu, F.; Zhao, X.; Liu, B.; Xu, J.; Yi, L.; Hu, S.; Wang, X.; Zuo, L.; et al. Urban expansion in China and its effect on cultivated land before and after initiating "Reform and Open Policy". *Sci. China (Earth Sci).* **2016**, *59*, 1930–1945. [CrossRef]
19. Xu, D.; Deng, X.; Guo, S.; Liu, S. Labor migration and farmland abandonment in rural China: Empirical results and policy implications. *J. Environ. Manag.* **2019**, *232*, 738–750. [CrossRef]
20. Huang, J.; Wang, X.; Zhi, H.; Huang, Z.; Rozelle, S. Subsidies and distortions in China's agriculture: Evidence from producer-level data. *Aust. J. Agric. Resour. Econ.* **2011**, *55*, 53–71. [CrossRef]
21. Wang, G.; Liu, Y.; Li, Y.; Chen, Y. Dynamic trends and driving forces of land use intensification of cultivated land in China. *J. Geogr. Sci.* **2015**, *25*, 45–57. [CrossRef]
22. Bai, Y.; Liu, M.; Yang, L. Calculation of Ecological Compensation Standards for Arable Land Based on the Value Flow of Support Services. *Land* **2021**, *10*, 719. [CrossRef]
23. Zhang, S.; Hu, W.; Li, M.; Guo, Z.; Wang, L.; Wu, L. Multiscale research on spatial supply-demand mismatches and synergic strategies of multifunctional cultivated land. *J. Environ. Manag.* **2021**, *299*, 113605. [CrossRef] [PubMed]
24. Liu, X.; Zhao, C.; Song, W. Review of the evolution of cultivated land protection policies in the period following China's reform and liberalization. *Land Use Policy* **2017**, *67*, 660–669. [CrossRef]
25. Claassen, R.; Cattaneo, A.; Johansson, R. Cost-effective design of agri-environmental payment programs: U.S. experience in theory and practice. *Ecol. Econ.* **2008**, *65*, 737–752. [CrossRef]
26. Tan, R.; Wang, R.; Sedlin, T. Land-Development Offset Policies in the Quest for Sustainability: What Can China Learn from Germany? *Sustainability* **2014**, *6*, 3400–3430. [CrossRef]
27. Baylis, K.; Peplow, S.; Rausser, G.; Simon, L. Agri-environmental policies in the EU and United States: A comparison. *Ecol. Econ.* **2008**, *65*, 753–764. [CrossRef]
28. Ma, S.; Swinton, S.M.; Lupi, F.; Jolejole-Foreman, C. Farmers' Willingness to Participate in Payment-for-Environmental-Services Programmes. *J. Agric. Econ.* **2012**, *63*, 604–626. [CrossRef]
29. Gardner, B.D. The Economics of Agricultural Land Preservation. *Am. J. Agric. Econ.* **1977**, *59*, 1027–1036. [CrossRef]
30. Corbelle-Rico, E.; Sánchez-Fernández, P.; López-Iglesias, E.; Lago-Peñas, S.; Da-Rocha, J. Putting land to work: An evaluation of the economic effects of recultivating abandoned farmland. *Land Use Policy* **2022**, *112*, 105808. [CrossRef]
31. Zhang, S.; Hu, W.; Zhang, J.; Li, M.; Zhu, Q. Mismatches in Suppliers' and Demanders' Cognition, Willingness and Behavior with Respect to Ecological Protection of Cultivated Land: Evidence from Caidian District, Wuhan, China. *Int. J. Environ. Res. Public Health* **2020**, *17*, 1156. [CrossRef]
32. Wang, N.; Zu, J.; Li, M.; Zhang, J.; Hao, J. Spatial Zoning of Cultivated Land in Shandong Province Based on the Trinity of Quantity, Quality and Ecology. *Sustainability* **2020**, *12*, 1849. [CrossRef]

33. Fischer, G.; Prieler, S.; van Velthuizen, H.; Lensink, S.M.; Londo, M.; de Wit, M. Biofuel production potentials in Europe: Sustainable use of cultivated land and pastures. Part I: Land productivity potentials. *Biomass Bioenergy* **2010**, *34*, 159–172. [CrossRef]
34. Wade, M.R.; Gurr, G.M.; Wratten, S.D. Ecological restoration of farmland: Progress and prospects. *Philos. Trans. R. Soc. Lond. Ser. B Biol. Sci.* **2008**, *363*, 831–847. [CrossRef]
35. Costanza, R.; D'Arge, R.; de Groot, R.; Farber, S.; Grasso, M.; Hannon, B.; Limburg, K.; Naeem, S.; O'Neill, R.V.; Paruelo, J.; et al. The value of the world's ecosystem services and natural capital. *Ecol. Econ.* **1998**, *25*, 3–15. [CrossRef]
36. Xie, G.; Lu, C.; Leng, Y.; Zheng, D.; Li, S. The Value Evaluation of Ecological Assets on the Qinghai-Tibet Plateau. *J. Nat. Resour.* **2003**, *2*, 189–196.
37. Xie, G.; Zhang, C.; Zhang, L.; Chen, W.; Li, S. Improvement of the Evaluation Method for Ecosystem Service Value Based on Per Unit Area. *J. Nat. Resour.* **2015**, *30*, 1243–1254.
38. Xie, H.; Huang, Y.; Choi, Y.; Shi, J. Evaluating the sustainable intensification of cultivated land use based on emergy analysis. *Technol. Forecast. Soc.* **2021**, *165*, 120449. [CrossRef]
39. Moreno-Sanchez, R.; Maldonado, J.H.; Wunder, S.; Borda-Almanza, C. Heterogeneous users and willingness to pay in an ongoing payment for watershed protection initiative in the Colombian Andes. *Ecol. Econ.* **2012**, *75*, 126–134. [CrossRef]
40. Xiong, K.; Kong, F. The Analysis of Farmers' Willingness to Accept and Its Influencing Factors for Ecological Compensation of Poyang Lake Wetland. *Procedia Eng.* **2017**, *174*, 835–842. [CrossRef]
41. Costanza, R.; de Groot, R.; Braat, L.; Kubiszewski, I.; Fioramonti, L.; Sutton, P.; Farber, S.; Grasso, M. Twenty years of ecosystem services: How far have we come and how far do we still need to go? *Ecosyst. Serv.* **2017**, *28*, 1–16. [CrossRef]
42. Zang, Z.; Zou, X.; Zuo, P.; Song, Q.; Wang, C.; Wang, J. Impact of landscape patterns on ecological vulnerability and ecosystem service values: An empirical analysis of Yancheng Nature Reserve in China. *Ecol. Indic.* **2017**, *72*, 142–152.
43. Jiang, C.H.; Li, G.Y.; Li, H.Q.; Li, M. The Study of Ecological Service Value of Farmland Ecosystem in the Beijing-Tianjin-Hebei Region. IOP conference series. *Earth Environ. Sci.* **2017**, *86*, 12004.
44. Shi, Y.; Duan, W.; Fleskens, L.; Li, M.; Hao, J. Study on evaluation of regional cultivated land quality based on resource-asset-capital attributes and its spatial mechanism. *Appl. Geogr.* **2020**, *125*, 102284. [CrossRef]
45. Buyanovsky, G.A.; Wagner, G.H. Carbon cycling in cultivated land and its global significance. *Global Chang. Biol.* **1998**, *4*, 131–141. [CrossRef]
46. Chen, L.; Hao, J.; Wang, F.; Yin, Y.; Gao, Y.; Duan, W.; Yang, J. Carbon sequestration function of cultivated land use system based on the carbon cycle for the Huang-Huai-Hai Plain. *Resour. Sci.* **2016**, *38*, 1039–1053.
47. Wang, J.; Cao, Y.; Fang, X.; Li, G.; Cao, Y. Identification of the trade-offs/synergies between rural landscape services in a spatially explicit way for sustainable rural development. *J. Environ. Manag.* **2021**, *300*, 113706. [CrossRef]
48. Zhang, F.; Wu, P.; Zhao, X.; Zhang, E.; Cheng, X.F. Effects of conservation tillage on soil water regimes and water use efficiency in farmland of Heihe River Basin in Northwest China. *Afr. J. Agric. Res.* **2011**, *6*, 5959–5966.
49. Song, W.; Deng, X. Effects of Urbanization-Induced Cultivated Land Loss on Ecosystem Services in the North China Plain. *Energies* **2015**, *8*, 5678–5693. [CrossRef]
50. Ma, B.; Ma, F.; Li, Z.; Wu, F. Effect of crops on rainfall redistribution processes under simulated rainfall. *Trans. Chin. Soc. Agric. Eng.* **2014**, *30*, 136–146.
51. Mamedov, A.I.; Levy, G.J. Soil erosion–runoff relations on cultivated land: Insights from laboratory studies. *Eur. J. Soil Sci.* **2019**, *70*, 686–696. [CrossRef]
52. Sun, X.; Xie, G.; Zhang, C.; Xiao, Y.; Lu, C. Services for Soil Conservation and Its Monetary Value of Chinese Cropping System. *J. Soil Water Conserv.* **2005**, *4*, 156–159.
53. Wu, R.; Long, Y.; Malanson, G.; Garber, P.; Zhang, S.; Li, D.; Zhao, P.; Wang, L.; Duo, H. Optimized Spatial Priorities for Biodiversity Conservation in China: A Systematic Conservation Planning Perspective. *PLoS ONE* **2014**, *9*, 103783.
54. Paracchini, M.L.; Zulian, G.; Kopperoinen, L.; Maes, J.; Schägner, J.P.; Termansen, M.; Zandersen, M.; Perez-Soba, M.; Scholefield, P.A.; Bidoglio, G. Mapping cultural ecosystem services: A framework to assess the potential for outdoor recreation across the EU. *Ecol. Indic.* **2014**, *45*, 371–385. [CrossRef]
55. Linkous, E.R. Transfer of development rights and urban land markets. *Environ. Plan. A* **2017**, *49*, 1122–1145. [CrossRef]
56. Yang, J.; Guo, A.; Li, Y.; Zhang, Y.; Li, X. Simulation of landscape spatial layout evolution in rural-urban fringe areas: A case study of Ganjingzi District. *Gisci. Remote Sens.* **2019**, *56*, 388–405. [CrossRef]
57. Li, C.; Gao, X.; Wu, J.; Wu, K. Demand prediction and regulation zoning of urban-industrial land: Evidence from Beijing-Tianjin-Hebei Urban Agglomeration, China. *Environ. Monit. Assess.* **2019**, *191*, 412. [CrossRef]
58. Liu, M.; Lu, S.; Song, Y.; Lei, L.; Hu, J.; Lv, W.; Zhou, W.; Cao, C.; Shi, H.; Yang, X.; et al. Microplastic and mesoplastic pollution in farmland soils in suburbs of Shanghai, China. *Environ. Pollut.* **2018**, *242*, 855–862. [CrossRef]

sustainability

MDPI

Article

Multi-Temporal Arable Land Monitoring in Arid Region of Northwest China Using a New Extraction Index

Xinyang Yu [1,2], Younggu Her [3], Xicun Zhu [1,*], Changhe Lu [2,4,*] and Xuefei Li [5]

1 College of Resources and Environment, Shandong Agricultural University, Tai'an 271018, China; yuxy.12b@igsnrr.ac.cn
2 Key Laboratory of Land Surface Pattern and Simulation, Chinese Academy of Sciences, Beijing 100101, China
3 Tropical Research and Education Center, Department of Agricultural and Biological Engineering, Institute of Food and Agricultural Sciences, University of Florida, Homestead, FL 33031, USA; yher@ufl.edu
4 College of Resources and Environment, University of Chinese Academy of Sciences, Beijing 100101, China
5 Linyi Natural Resources Development Service Center, Linyi 276000, China; lixuefeilinyi@163.com
* Correspondence: zxc@sdau.edu.cn (X.Z.); luch@igsnrr.ac.cn (C.L.)

Abstract: Development of a high-accuracy method to extract arable land using effective data sources is crucial to detect and monitor arable land dynamics, servicing land protection and sustainable development. In this study, a new arable land extraction index (ALEI) based on spectral analysis was proposed, examined by ground truth data, and then applied to the Hexi Corridor in northwest China. The arable land and its change patterns during 1990–2020 were extracted and identified using 40 Landsat TM/OLI images acquired in 1990, 2000, 2010, and 2020. The results demonstrated that the proposed method can distinguish arable land areas accurately, with the User's (Producer's) accuracy and overall accuracy (kappa coefficient) exceeding 0.90 (0.88) and 0.89 (0.87), respectively. The mean relative error calculated using field survey data obtained in 2012 and 2020 was 0.169 and 0.191, respectively, indicating the feasibility of the ALEI method in arable land extracting. The study found that arable land area in the Hexi Corridor was 13217.58 km^2 in 2020, significantly increased by 25.33% compared to that in 1990. At 10-year intervals, the arable land experienced different change patterns. The study results indicate that ALEI index is a promising tool used to effectively extract arable land in the arid area.

Keywords: arable land extraction index; arid region; Landsat image; Infrared band; Shortwave band; Hexi Corridor

Citation: Yu, X.; Her, Y.; Zhu, X.; Lu, C.; Li, X. Multi-Temporal Arable Land Monitoring in Arid Region of Northwest China Using a New Extraction Index. *Sustainability* **2021**, *13*, 5274. https://doi.org/10.3390/su13095274

Academic Editor: Yu Cao

Received: 14 March 2021
Accepted: 28 April 2021
Published: 8 May 2021

1. Introduction

Human-induced land use/cover change has caused land development (e.g., arable land, urban land), expanding and encroaching into natural areas [1,2]. Arable land in arid regions is mainly distributed in artificial oases. Compared to other land use/cover types in an arid region, e.g., woodland, arable land is more vulnerable to extreme weather and human disturbance [3,4]. Accurately detecting arable land areas and monitoring its dynamic change is indispensible for the development of land management policies to ensure sustainable development.

Early land use/cover change studies have utilized historical records and hand-drawn maps to depict arable land in arid regions [5,6]. Since the 1990s, land use/cover maps from remotely sensed images have been employed as the main data source to monitor the arable land's status and its response to external disturbance [7–9], because remotely sensed images can help retrieve the context, direction, and rate of arable land change in a time- and labor-saving way [10,11]. According to the spatial resolution, remotely sensed images can be categorized to coarse-, middle- and high-resolution classes. High-resolution (<10 m) images are expensive and the imaging scan width is relatively small, and thus are generally not feasible for large-scope land use/cover monitoring (e.g., >10 km^2). Coarse-resolution images, such as MODIS (250 m/500 m/1 km) and AVHRR (1 km) series data set, have been widely used to study the land use/cover change at global and regional levels [12–14]. However, due to their coarse resolution, pixels are more likely to contain different land use/cover types, which are referred to as mixed pixels. Compared to pure pixel (a pixel that contains only one land use/cover type), a mixed pixel in the coarse-resolution images can lead to low accuracy of arable land detection. Given the high demand of precise land use/cover monitoring, the development of cloud computing platforms such as Google Earth Engine, and efficient methods to process images [15,16], middle-resolution Landsat series images, as first launched in 1972, have gradually became one of the most practical data sources to delineate and monitor the dynamics of land surface change at a regional scale (e.g., [17–19]). Compared to the coarse-resolution images and other middle-resolution images, Landsat images can identify land use/cover at low cost and acceptable accuracy [20–22] and help monitor the dynamic change of land uses/cover in longer time intervals (1972 to date).

There are volumes and varieties of Landsat data available, however, methods used to delineate arable land at a regional scale suffered development lag. For a long period, a high proportion of information from remotely sensed images was distinguished by visual interpretation, which is labor intensive and time consuming [23]. To improve the efficiency of arable land identification and minimize the artificial errors, vegetation indices, such as Normalized Difference Vegetation Index (NDVI), were employed by studies to differentiate arable land areas [24]. For instance, Xie et al. (2018) [25] used the NDVI threshold method for vegetation extraction and monitored the spatio-temporal changes of oases at a regional scale and found that different NDVI values can help identify the boundary of oases. Venkatappa et al. (2019) [26] developed a robust phenology-based classification method to accurately determine the threshold values of vegetation types.

Vegetation Index-based method is easy to delineate the boundary of arable land and desert, for they show apparently different patterns in the Infrared and Red spectral bands. Recently, the open, shared data sets of Globeland30 that fulfilled the 10-year-interval land use/cover classification were released for free download [27]. However, their classification accuracy of arable land in northwest China is 0.8315 [28], which can be further improved. Besides, in the agro-pastoral zones of arid region, vegetation index values of arable land and other vegetation types (woodland, shaded vegetation, etc.) share range overlap due to their similar spectral change patterns, and thus the vegetation index threshold method may not be sufficient to distinguish arable land in the complicated agro-pastoral transition zones. Currently used classification methods may misclassify the woodland and shaded vegetation to arable land, resulting in more bias and uncertainties of arable land extraction in arid region. Therefore, they need a feasible algorithm to extract arable land to ensure accurate detection of arable land areas in an arid region.

To this end, this study proposed a framework to efficiently distinguish arable land area and its dynamic change patterns. Multi-temporal Landsat images and different sources of auxiliary data were collected and processed to examine the performance of the proposed framework. The method can be used for arable land monitoring in an arid region, and the results can be used as a reference for the agricultural development in the Hexi Corridor region.

2. Materials and Methods

2.1. Study Area

The Hexi Corridor (90°11′–101°29′ E, 37°14′–42°15′ N) is a belt region located in northwest China. The altitude of the study area ranges from 790 to 5780 m and the climate is arid, temperate monsoon, characterized by a hot summer and cold winter. The annual average temperature during the study period ranged from 4 to 10 °C, and the annual average precipitation was from 100 to 400 mm in the region, showing a decreasing trend from southeast to northwest. The region is comprised of three inland rivers, all originated from the Qinghai-Tibet Plateau, including the Shiyanghe river basin, the Heihe river basin, and the Shulehe river basin, distributing from southeast to northwest (Figure 1).

Arable land in the study area is mainly distributed in the oases of the Jinchang, Minqin, Sunan, Wuwei, and Yongchang administrative counties in the Shiyanghe river basin; in Gaotai, Jiayuguan, Jinta, Jiuquan, Linze, Minle, Shandan, and Zhangye administrative counties in the Heihe river basin; and in Akesai, Dunhuang, Guazhou, Subei, and Yumen administrative counties in the Shulehe river basin. By thousands of years of cultivation, the arable land has differed clearly from other land use types, so we selected this region to examine the method for arable land extracting.

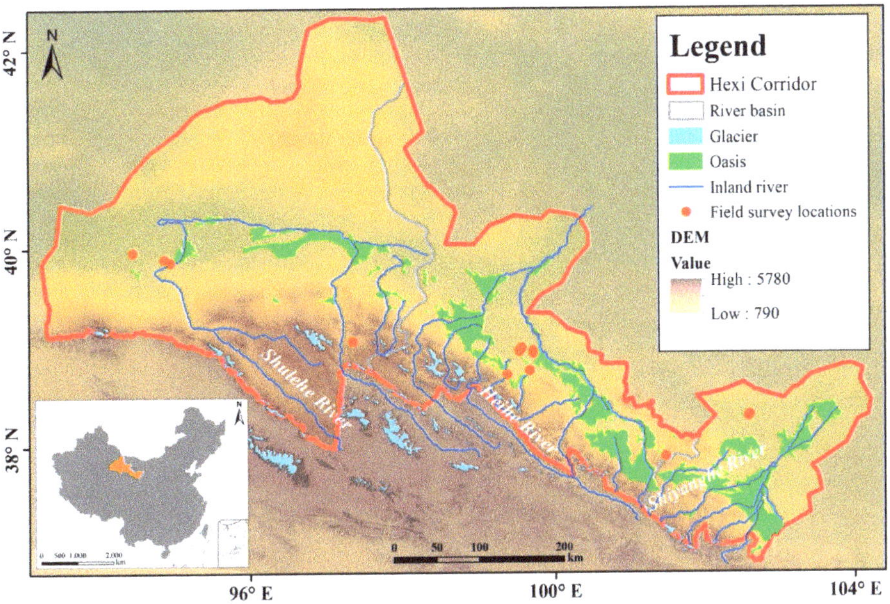

Figure 1. Location of the study area.

2.2. General Prodcedure of Data Processing

The arable land was extracted in four successive steps (Figure 2): (1) data selection and processing, (2) pure pixels' selection, (3) index framework setup, and (4) arable land extracting and accuracy estimation.

2.2.1. Data Sources and Preprocessing

A total of 40 Landsat images (10 images for each year, see Appendix A) were collected for the years 1990, 2000, 2010, and 2020. The Landsat-series images covered seven paths from east to west (131–137) and three rows from north to south (032–034) and contained 30 Landsat-5 TM and 10 Landsat-8 *OLI* images acquired in the summer season. The Google Earth Engine [21,29] was used to identify the images that were cloud-free and possessed the highest crop coverage rates through the years. Thus, 40 raw format images acquired from 16 June to 6 September during the 30 years were selected and downloaded from the public domain of the Earth Resources Observatory and Science Center (EROS, http://eros.usgs.gov, accessed on 18 October 2020) using a Python bulk download application. The images were then rectified to the Universal Transverse Mercator (*UTM*) projection using at least 45 well-distributed ground control points (GCPs) and nearest neighborhood resampling method for each image, and the root mean square error (*RMSE*) was set to be less than 1 pixel. Radiometric calibration, fast line-of-sight atmospheric analysis of spectral hypercubes (*FLAASH*) atmospheric correction, and geometry correction were conducted to obtain surface reflectance images using a batch-processing IDL program.

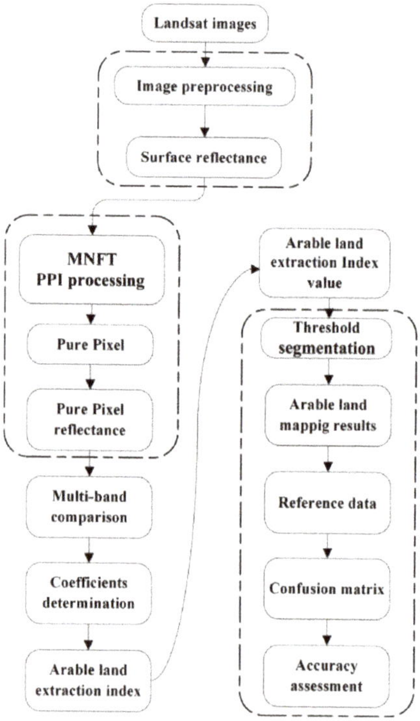

Figure 2. Workflow chart of arable land extraction.

2.2.2. Selection of Pure Pixels

Owing to the high separability characteristics between vegetation (arable land, woodland, etc.) and other land use/cover types (desert, urban and rural areas, glacier, waterbody, etc.), the NDVI threshold method can be used to distinguish and locate vegetation areas in the remotely sensed images. However, due to apparent overlaps in the reflectance and impact of terrain fluctuation, arable land is hard to be accurately separated from woodland and shaded vegetation based on NDVI differences (Figure 3). Here the three land use/cover types (arable land, woodland, and shaded vegetation) were visually interpreted from five randomly selected Landsat images for each year with the aid of Google Earth Pro images, and their *NDVI* values were calculated using the equation below.

$$NDVI = Float(NIR - R)/\ Float\ (NIR + R) \quad (1)$$

where *NIR* represents the reflectance value of Near-infrared band, and *R* is the reflectance value of Red band.

The Minimum Noise Fraction Transform (MNFT) and Pixel Purity Index (PPI) were conducted on the Landsat images to find pure pixels, which were used to detect cues of spectral difference between arable land and other vegetation types [30,31]. At first, the MNFT was used to eliminate the noise in the images by principal component transformations in two steps. The first transformation (based on the estimated noise covariance matrix) was to separate and re-adjust the noise in the data, by minimizing the variance of the transformed noise data without correlation between bands. The second step was to transform the standard principal component of the noise-whitened data [32].

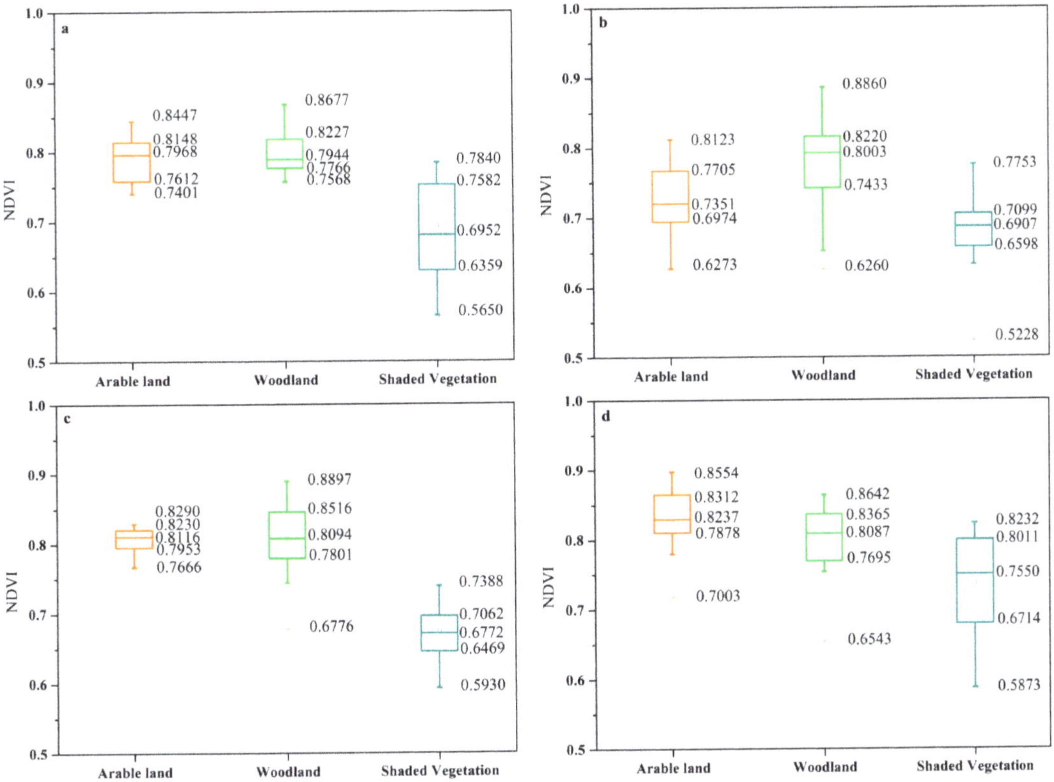

Figure 3. NDVI distributions of arable land, woodland, and shaded vegetation in the Landsat images in the year 1990 (**a**), 2000 (**b**), 2010 (**c**), and 2020 (**d**). The three land use/cover types were visually interpreted from the randomly selected Landsat images with the aid of Google Earth Pro images. Each box plot explains the location of the 0th, 25th, 50th, 75th, and 100th percentiles using horizontal lines (boxes and whiskers).

The pixel purity index (PPI) has been widely used in multi-spectral image analysis for pure pixel detection. After the MNFT processing, a set of k unit vectors $\{skewer_j\}_{j=1}^{k}$ was generated, where k is a sufficiently large, positive integer. For each $skewer_j$, all the data sample vectors were projected onto $skewer_j$ to find sample vectors at its extreme positions to form an extrema set for this particular $skewer_j$, denoted by $S_{extrema}(skewer_j)$. Despite the fact that a different $skewer_j$ generates a different extrema set $S_{extrema}(skewer_j)$, it is likely that some sample vectors may appear in more than one extrema set. Define an indicator function of a set S, $I_S(r)$ by

$$I_S(r) = \begin{cases} 1, & if\ r \in S \\ 0, & if\ r \notin S \end{cases} \text{ and} \qquad (2)$$
$$N_{PPI}(r) = \sum_j I_{Sextrema(skewer_j^k)}(r)$$

where $N_{PPI}(r)$ is defined to be the PPI score of sample vector r [33]. A large PPI value (e.g., 1) indicates a high possibility of a pure pixel [34].

The pixels with high PPI values were then exported to establish a new region of interest (ROI) and then linked to the Google Earth Pro images (0.5-m spatial resolution) to further examine the purity of these pixels. Finally, about 400 pure pixel samples were selected in each image for further analysis.

Figure 4 shows the reflectance of the six bands (Blue, Green, Red, Infrared, shortwave 1, and shortwave 2) of the p132r34 image as an example. Comparison analysis found that the spectral change patterns of arable land, woodland, and shaded vegetation were similar to each other. However, the value of the Infrared band was apparently different from other bands and, thus, this band was selected as the basis to design the framework to distinguish arable land areas.

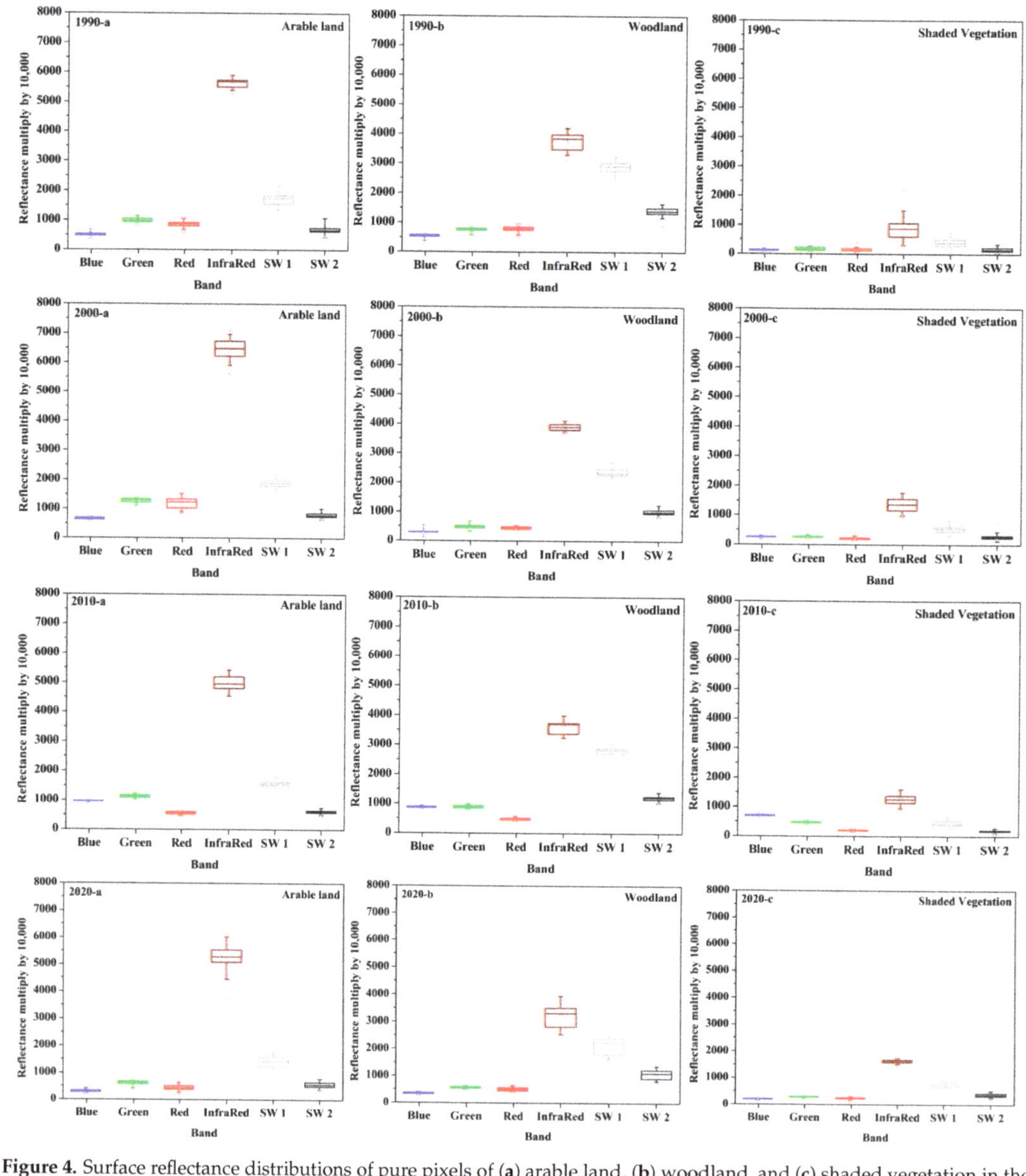

Figure 4. Surface reflectance distributions of pure pixels of (**a**) arable land, (**b**) woodland, and (**c**) shaded vegetation in the example image. The reflectance value was multiplied by 10,000. Each box plot explains the location of the 0th, 25th, 50th, 75th, and 100th percentiles using horizontal lines (boxes and whiskers).

2.2.3. Formulation of Arable Land Extraction Index (ALEI)

Spectral difference is an important foundation to discern arable land. As the wavelength increasing from visible to the shortwave bands, the reflectance of arable land and natural vegetation showed the same change pattern. They increased in the Green band, slightly decreased in the Red band, significantly increased in the Infrared band, and then decreased in the Shortwave bands (Figure 4). The shaded vegetation shared the same change pattern with that of woodland but the reflectance in the Infrared was much less than those of arable land and woodland (Figure 4) and the other images shared the same spectral distribution characteristics. This can help separate shaded vegetation and eliminate its impact on arable land extract. Therefore, the main issue was to separate arable land from woodland in the agro-pastoral zones. After analyzing the spectral change patterns of arable land and woodland, the Infrared and Shortwave 1 were selected to form an arable land extraction index to maximize the contrast between arable land and woodland.

$$ALEI = \rho_{Infrared} - \omega * \rho_{Shortwave\ 1} \quad (3)$$

where $\rho_{Infrared}$ is the reflectance value of the Infrared band and $\rho_{Shortwave\ 1}$ is the reflectance value of the first shortwave band.

For successive identification of arable land by excluding other easily confused land use/cover types, the key is to determine the coefficient ω to maximize the difference between arable land and natural vegetation. So, iteration computations with different ω values were conducted, and the results are presented in Figure 5b. Considering the image acquisition date and land use/cover distribution in the study area, the value cannot be too small or too large, as such may lower the desired difference to separate arable land and woodland. For instance, if ω were 1.4, the index values for the two land use/cover types are positive, and when the ω value increases, e.g., to larger than 2.2, their values become negative. By iterative examinations, the suitable ω value can be set to 1.6 or 2.0 for the formula, but they both fail to fulfill the purpose of maximizing the difference between arable land and woodland. By assigning the coefficient ω to be 1.8, the pixel with positive index value represents arable land and the pixel-processing negative one is woodland. Here, we consented that 0 was a default threshold for separating arable land from woodland, because pixels with positive and negative ALEI values are easier to be classified into different land use/cover categories.

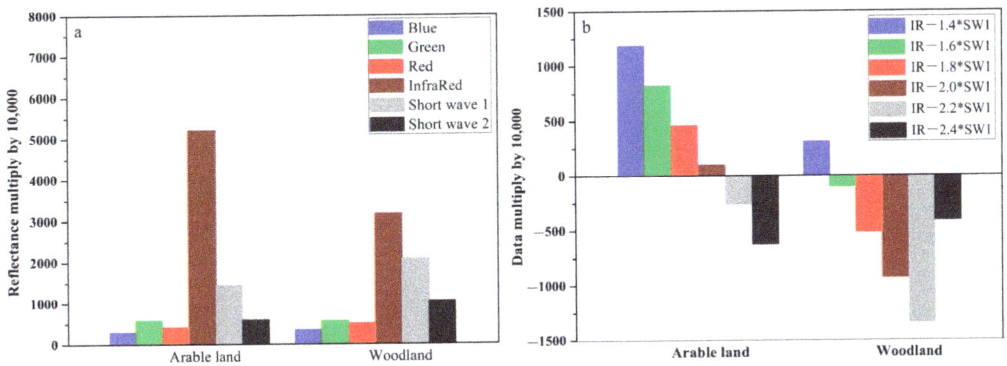

Figure 5. Process of arable land extraction index formation. (**a**) Average reflectance of the six bands in the Landsat images; (**b**) a process of determining the optimal coefficients of the proposed index.

2.2.4. Verification

After the determination of ω parameter, the ALEI was applied to extract the arable land. The arable land extraction framework was built using a two-branch decision tree method. At first, "B1 LT 2500" was used to eliminate the impact of shaded vegetation to arable land extraction, where "B1" is the abbreviation of Band 1, i.e., the Infrared bands of Landsat image, and "LT" represents "Less Than", i.e., the pixels in the Landsat images with Infrared value of less than 0.25 that were classified to shaded vegetation. Secondly, the remaining pixels were further processed using the "B2 GT 1.8" to distinguish arable land, in which the "B2" (Band 2) is the ALEI-based results calculated using Equation (3) and "GT" is the abbreviation of "Greater Than". The pixels with ALEI values greater than 1.8 were grouped into the arable land class. In this study, 40 decision trees were established separately and 40 different B1 and B2 sets were used to obtain the computation results.

A 160,000-pixel ground truth data set was randomly selected, and the land use types for each image were visually interpreted based on Google Earth Pro online images (0.5-m spatial resolution). With this ground truth data set, the currency of the ALEI-based results was examined using four coefficients, i.e., user's accuracy, producer's accuracy, overall accuracy, and Kappa coefficient [35]. The user's accuracy is the ratio of the total number of pixels correctly classified into class i (diagonal value in the confusion matrix) and the total number of pixels classified into class i by the classifier (the sum of class i rows in the confusion matrix). The producer's accuracy is the ratio of the number of pixels (diagonal value in the confusion matrix) that the classifier correctly classifies for the entire image into class i and the true total number of class i (the sum of class i columns in the confusion matrix).

Furthermore, intensive field surveys were conducted for 15 arable land areas in July 2012 and July 2020, for each of which the arable land area boundary was recorded by walking around it using a high-precision tracking GPS (position dilution of precision ≤ 4, error < 10 m). The recorded boundary for each of the 15 arable land areas was input into ArcGIS 10.6, and then the area was computed. These data were then used for in situ verification of the arable land area obtained from the ALEI method in 2010 and 2020, with the mean relative error (*MRE*) as below.

$$MRE = \frac{\sum_{i=1}^{n}(abs(x_i - y_i)/y_i)}{n} \tag{4}$$

where x_i is the ith area obtained from field survey, y_i is the ith area from ALEI method, and n is 15, the total number of the field survey locations.

2.3. Dynamic Change Monitoring of Arable Land

Arable land dynamic change during 1990–2000, 2000–2010, and 2010–2020 in the three river basins was computed by the equations below:

$$\vec{R} = \left(\sqrt[t]{\frac{A_b}{A_a}} - 1\right) * 100\% \tag{5}$$

$$T_s = \frac{\Delta A_{gain} - \Delta A_{loss}}{\Delta A_{gain} + \Delta A_{loss}},\ \Delta A_{gain} + \Delta A_{loss} \neq 0,\ T_s \in [-1, 1] \tag{6}$$

where $\vec{R}$ is the net change rate index and T_s is the status and change trend index. A_a and A_b are the total area of arable land in the initial and end year of the research period, ΔA_{gain} and ΔA_{loss} represent the area gained and lost during 1990–2020, respectively, and t is the time duration of the time interval of 10 years. From the Equations (5) and (6), the higher the $\vec{R}$ is, the more the overall growth is. When T_s is less than 0, the total arable land is reduced; otherwise, it is increased.

3. Results and Discussion

3.1. Verification of Arable Land Extraction

The overall accuracy (kappa coefficient) values of arable land extraction using the ALEI method all exceeded 0.89 (0.87), and both user's and producer's accuracies were higher than 0.84, indicating that the classification results were feasible for arable land change analysis (Table 1). Four 160,000-pixel zones were randomly selected in the images of p131r34, p132r34, p133r33, and p135r32 from southeast to northwest to visually show the extraction results. The original Landsat images, the arable land results using the ALEI and the NDVI threshold method, are shown in Figure 6. Both the ALEI and NDVI methods can distinguish vegetation from non-vegetation land use/cover types such as cloud, waterbody, riverbed, urban, and rural areas. However, NDVI-based method confused arable land areas with woodland and shaded vegetation (Figure 6f,l), which resulted in overestimation of arable land areas.

Table 1. Accuracy assessment of arable land and other vegetation types.

Land Cover	Accuracy Type	Year			
		1990	2000	2010	2020
Arable land	User's Accuracy	0.90–0.92	0.92–0.94	0.92–0.95	0.93–0.94
	Producer's Accuracy	0.88–0.91	0.91–0.94	0.90–0.92	0.92–0.93
Woodland	User's Accuracy	0.87–0.90	0.89–0.93	0.89–0.91	0.89–0.93
	Producer's Accuracy	0.85–0.89	0.86–0.91	0.87–0.90	0.88–0.91
Shaded vegetation	User's Accuracy	0.90–0.93	0.92–0.95	0.90–0.92	0.92–0.94
	Producer's Accuracy	0.90–0.93	0.90–0.93	0.89–0.92	0.89–0.91
All	Overall Accuracy	0.89–0.93	0.91–0.94	0.91–0.94	0.90–0.93
All	Kappa Coefficient	0.87–0.90	0.89–0.91	0.90–0.92	0.89–0.91

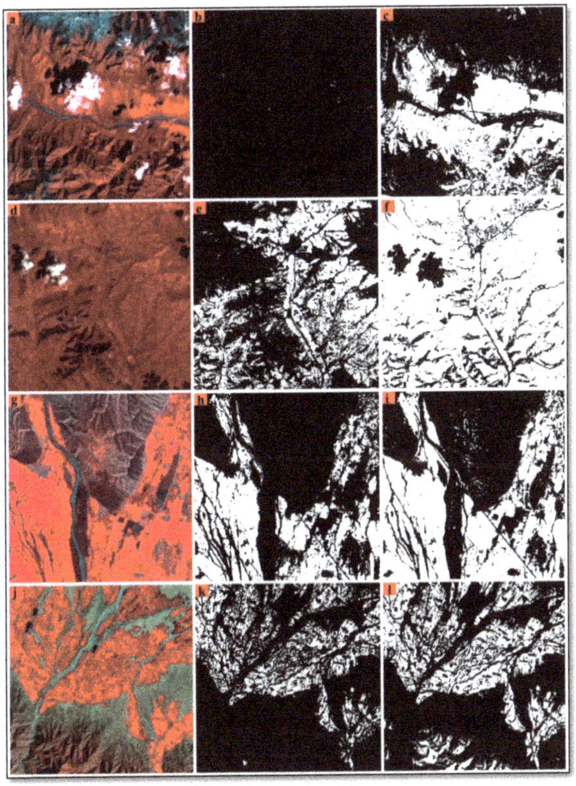

Figure 6. Extraction results of arable land in representative agro-pastoral transition zones. (**a**) The p131r34 image acquired in 1990; (**d**) p132r34 image in 2000; (**g**) p133r33 image in 2010; (**j**) p135r32 image in 2020. Result map of (**b**,**e**,**h**,**k**): the extraction results based on the proposed ALEI method in this study; (**c**,**f**,**i**,**l**) the arable land areas from the NDVI threshold method. White areas represent extracted arable land and black ones are the other land cover types.

The accuracy of arable land extraction was further verified using areas of arable land obtained in the field survey in the summer season of 2012 and 2020. The correlation between field survey data and ALEI-based data are shown in Figure 7. Both the areas of arable land in 2010 and 2020 showed a high correlation, the coefficients of the years 2010 and 2020 were 0.969 and 0.963, and the *MRE* values were 0.169 and 0.191, respectively. However, the ALEI results were slightly overestimated compared to the field survey data. This was mainly attributed to the mixing pixels existing in these Landsat images. For instance, a pixel that contained 80% arable land and 20% other land use/cover types was classified as arable land using the ALEI method, resulting in some error. Overall, the classification accuracy using the ALEI method was acceptable and the results can be used to monitor the dynamic change of arable land in the study area.

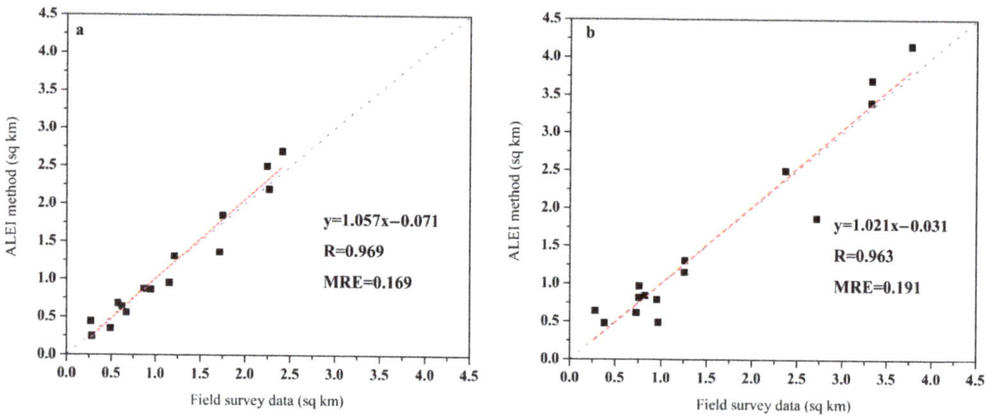

Figure 7. Estimates of arable land using the ALEI method vs. field survey data obtained in 2010 (**a**) and 2020 (**b**).

3.2. Change Pattern of Arable Land in the Study Area

The extracted arable land in the study area during 1990–2020 are presented in Figure 8. In 1990, the total area of arable land was 10,546.33 km^2, of which 34.45%, 46.42%, and 19.13% was distributed in the Shiyanghe, the Heihe, and the Shulehe river basins, respectively. In 2020, the total arable land area was 13,217.58 km^2, increased by 25.33% compared to that in 1990, which reached its extreme value. During the three 10-year study intervals, arable land in the Hexi Corridor showed a continually increasing trend with the increase rates of 4.99%, 10.86%, and 7.43%, respectively. The most significant increase occurred during 2000–2010.

In the three inland river basins, arable land showed a significantly expanding trend during the whole study period, while they showed different change patterns in the three 10-year intervals:

1. In the Shiyanghe river basin, arable land area increased rapidly with an annual rate of 2.36 % from 1990 to 2020, which was the lowest among the three inland river basins (Heihe river basin 2.60%/year, Shulehe river basin 2.68%/year). In this river basin, the most rapid arable land expansion occurred during 1990–2000, with a net change rate $\vec{R}$ of 0.99%. The trend index T_s of this 10-year interval was 0.51. It had not much difference with the other two 10-year periods (0.53 during 2000–2010 and 0.50 in 2010–2020 interval, Table 2). This indicated that, though arable land in the Shiyanghe basin expanded prominently, the area lost was also significant (Figure 9).

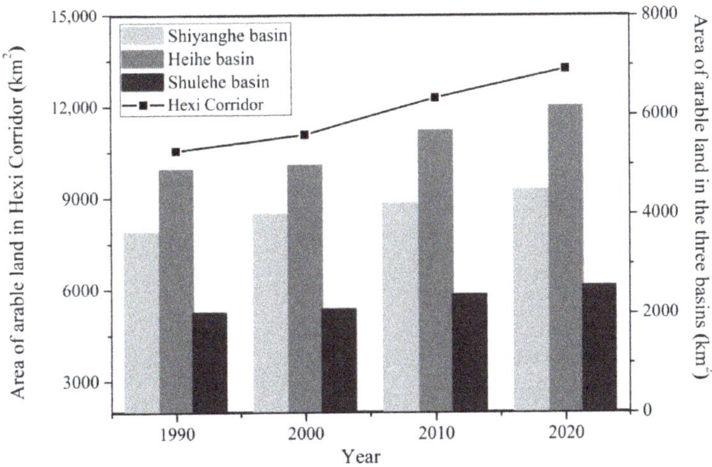

Figure 8. Oasis change in the Hexi Corridor during 1990–2020. The black, block line shows the change trend of arable land area, and the bars indicate the arable land area of the Shiyanghe, Heihe, and Shulehe inland river basins.

Table 2. Change of arable land in the three 10-year intervals.

River Basin	Net Change Rate ($\vec{R}$, %)/Status and Trend Index (T_s)		
	1st Interval	2nd Interval	3rd Interval
Shiyanghe	0.99/0.51	0.51/0.53	0.63/0.50
Heihe	0.17/0.18	1.33/0.82	0.83/0.78
Shulehe	0.33/0.23	1.32/0.69	0.74/0.52

2. Arable land in the Heihe river basin during 1990–2000 extended slightly in area (0.17%/year), and the peak cumulative period of arable land occurred in the second 10-year interval (2000–2010, Figure 9). The area of arable land went up to 5681.04 km^2 with the net change rate $\vec{R}$ of 1.33 % and trend index T_s of 0.82, both of which were the highest among the three river basins. During 2010–2020, arable land in the Heihe river basin accumulated another 488.15 km^2, and the net change rate $\vec{R}$ and trend index T_s showed a decrease trend compared to those in the second 10-year interval (Table 2).
3. Arable land area of the Shulehe river basin in 1990 (2017.87 km2) only accounted for 55.55% and 41.22 % of that in the Shiyanghe and Heihe basins, respectively. However, arable land in the Shulehe river basin showed a similar change trend with that in the Heihe river basin (Figure 8 and Table 2). During 2000–2010, the area of arable land in the Shulehe basin increased by 292.12 km2 (14.09 %) and during 2010–2020, it further expanded by 7.61 % (Figure 9).

3.3. Discussion

This study established a framework to distinguish arable land and its change patterns in the Hexi Corridor, northwest China. The ALEI method was proposed to distinguish arable land based on 40 Landsat series images. The results demonstrated that the ALEI method is a feasible tool for extracting arable land areas accurately. The arable land in the study area showed an ongoing, expanding trend during 1990–2020, while the three inland river basins experienced different change patterns among the three 10-year intervals.

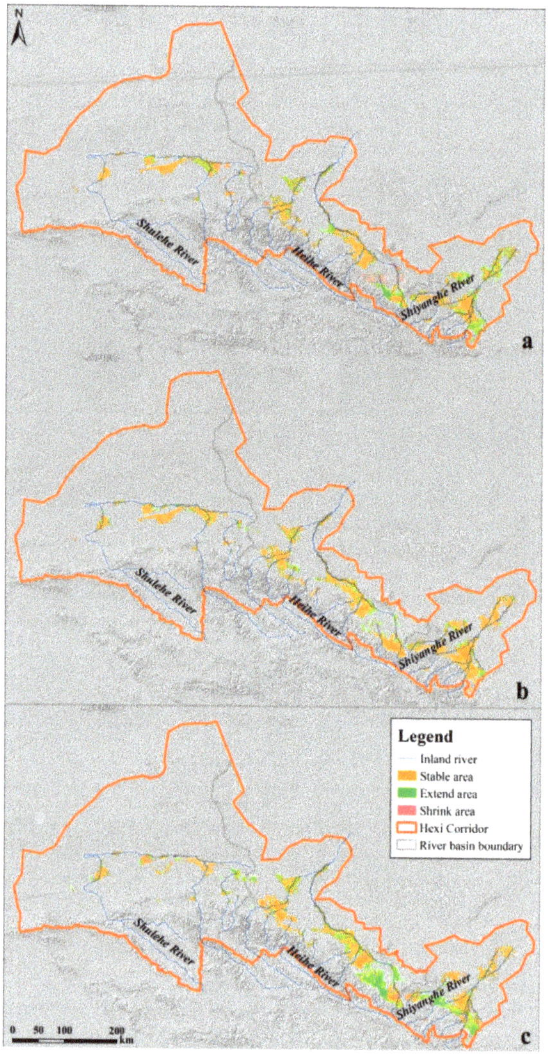

Figure 9. Arable land change in the study area during the three 10-year intervals. (**a**) 1990–2000; (**b**) 2000–2010; (**c**) 2010–2020.

Compared to the NDVI threshold method, ALEI can help distinguish arable land in the study area more accurately, especially in the agro-pastoral zones. The verification results demonstrated that the producer's accuracy of arable land extraction using the ALEI method ranged from 0.88 to 0.94, which increased by 6.02% to 13.25% compared to that of the Globeland30 (0.8315) [28]. The study results revealed the status and change patterns of arable land in the Hexi Corridor and can be used as references for decision makers to implement more targeted land protection policies and regulations, to balance the socioeconomic and ecological conflicts.

The ALEI method uses an efficient process framework in which the works can be all or semi-automatically conducted, i.e., ALEI can be computed with a batch-processing module automatically, after the Landsat images are downloaded using a bulk download application. Overall, the proposed methodology is understandable and easy to use. As artificial oases in arid regions share similar vegetation types and Landsat image is one of the most popular remotely sensed data sets covering most of the arid regions worldwide [36], ALEI is a promising tool to accurately extract arable land for land use planning in other arid regions. However, there are some limitations that were noted. Firstly, the field surveys for the verification was limited, covering an area of less than 4 km^2, because large arable land areas are hard to be measured using the "walking-GPS" way. Recently, an unmanned aerial system (UAS) was applied in the remote sensing studies [37,38] to measure large, arable land areas as the ground truth data. Considering that the field data in 2012 were collected by the "walking-GPS", so the data in 2020 were also collected by the same way to maintain the consistency and minimize the uncertainties. Secondly, the ALEI method was used and tested in the Hexi Corridor, and more applications in other arid regions could be done to further examine its usability and accuracy.

4. Conclusions

This study developed, examined, and applied a framework and an index of ALEI to identify arable land and its change patterns during past decades in a representative arid region. The results demonstrated that the proposed framework can help identify arable land areas accurately and found that the arable land in the study area extended significantly during 1990–2020. The findings of this study offer essential data to support the formulation and implementation of arable land protection and development policies.

Author Contributions: Conceptualization, X.Y. and X.Z.; data analysis, X.Y. and Y.H.; investigation, X.L.; methodology, Y.H.; writing—original draft, X.Y.; review and editing, Y.H., C.L. and X.Z. All authors have read and agreed to the published version of the manuscript.

Funding: This research was funded by the Open fund of Key Laboratory of Land Surface Pattern and Simulation, Chinese Academy of Sciences, grant number LBKF201802. The APC was funded by the Open fund of Key Laboratory of Land Surface Pattern and Simulation, Chinese Academy of Sciences, grant number LBKF201802. The funders had no role in the study design, data collection and analysis, decision to publish, or preparation of the article.

Institutional Review Board Statement: Not applicable.

Data Availability Statement: The data used to support the findings of this study are available from the corresponding author upon request.

Acknowledgments: The authors would like to thank the editors and reviewers for helpful comments.

Conflicts of Interest: The authors declare no conflict of interest.

Appendix A

Table A1. Metadata of Landsat images covering the oases in the study area.

Landsat ID	1990		2000		2010		2020	
	Sensor	Acquisition Date	Sensor	Acquisition Date	Sensor	Acquisition Date	Sensor	Acquisition Date
p131r33	TM	22 June 1990	TM	19 July 2000	TM	15 July 2010	OLI	10 July 2020
p132r33	TM	1 September 1990	TM	11 August 2000	TM	23 August 2010	OLI	17 July 2020
p132r34	TM	1 September 1990	TM	11 August 2000	TM	8 September 2010	OLI	17 July 2020
p133r33	TM	23 August 1990	TM	18 August 2000	TM	14 August 2010	OLI	9 August 2020
p134r32	TM	30 August 1990	TM	8 July 2000	TM	21 August 2010	OLI	1 September 2020
p134r33	TM	30 August 1990	TM	8 July 2000	TM	21 August 2010	OLI	1 September 2020
p135r32	TM	21 August 1990	TM	29 July 2000	TM	27 July 2010	OLI	23 August 2020
p136r32	TM	28 August 1990	TM	20 June 2000	TM	16 June 2010	OLI	30 August 2020
P136r33	TM	28 August 1990	TM	20 June 2000	TM	16 June 2010	OLI	30 August 2020
p137r32	TM	19 August 1990	TM	13 July 2000	TM	9 July 2010	OLI	6 September 2020

References

1. Chen, B.; Xiao, X.; Li, X.; Pan, L.; Doughty, R.; Ma, J.; Dong, J.; Qin, Y.; Zhao, B.; Wu, Z.; et al. A mangrove forest map of China in 2015: Analysis of time series Landsat 7/8 and Sentinel-1A imagery in Google Earth Engine cloud computing platform. *ISPRS J. Photogramm. Remote Sens.* **2017**, *131*, 104–120. [CrossRef]
2. Meyer, W.B.; Turner, B.L. Human Population Growth and Global Land-Use/Cover Change. *Annu. Rev. Ecol. Syst.* **1992**, *23*, 39–61. [CrossRef]

3. Guo, Y.; Shen, Y. Agricultural water supply/demand changes under projected future climate change in the arid region of northwestern China. *J. Hydrol.* **2016**, *540*, 257–273. [CrossRef]
4. Manjaribe, C.; Frasier, C.L.; Rakouth, B.; Louis, E.E. Ecological Restoration and Reforestation of Fragmented Forests in Kianjavato, Madagascar. *Int. J. Ecol.* **2013**, *2013*, 1–12. [CrossRef]
5. Li, X.; Shao, M.; Zhao, C.; Jia, X. Spatial variability of soil water content and related factors across the Hexi Corridor of China. *J. Arid. Land* **2019**, *11*, 123–134. [CrossRef]
6. Liu, F.; Yang, Y.; Shi, Z.; Storozum, M.J.; Dong, G. Human settlement and wood utilization along the mainstream of Heihe River basin, northwest China in historical period. *Quat. Int.* **2019**, *516*, 141–148. [CrossRef]
7. Gloaguen, R.; Goerner, A.; Makeschin, F. Monitoring of the Ecuadorian mountain rainforest with remote sensing. *J. Appl. Remote Sens.* **2007**, *1*, 013527. [CrossRef]
8. Li, Y.; Ge, Q.; Wang, H.; Liu, H.; Tao, Z. Relationships between climate change, agricultural development and social stability in the Hexi Corridor over the last 2000 years. *Sci. China Earth Sci.* **2019**, *62*, 1453–1460. [CrossRef]
9. Ma, L.; Cheng, W.; Qi, J. Coordinated evaluation and development model of oasis urbanization from the perspective of new urbanization: A case study in Shandan County of Hexi Corridor, China. *Sustain. Cities Soc.* **2018**, *39*, 78–92. [CrossRef]
10. Benkhattab, F.Z.; Hakkou, M.; Bagdanavičiūtė, I.; El Mrini, A.; Zagaoui, H.; Rhinane, H.; Maanan, M. Spatial–temporal analysis of the shoreline change rate using automatic computation and geospatial tools along the Tetouan coast in Morocco. *Nat. Hazards* **2020**, *104*, 1–18. [CrossRef]
11. Ku, C.-A. Exploring the Spatial and Temporal Relationship between Air Quality and Urban Land-Use Patterns Based on an Integrated Method. *Sustainability* **2020**, *12*, 2964. [CrossRef]
12. Friedl, M.A.; McIver, D.K.; Baccini, A.; Gao, F.; Schaaf, C.; Hodges, J.C.F.; Zhang, X.Y.; Muchoney, D.; Strahler, A.H.; Woodcock, C.E.; et al. Global land cover mapping from MODIS: Algorithms and early results. *Remote Sens. Environ.* **2002**, *83*, 287–302. [CrossRef]
13. Kibret, K.S.; Marohn, C.; Cadisch, G. Use of MODIS EVI to map crop phenology, identify cropping systems, detect land use change and drought risk in Ethiopia–an application of Google Earth Engine. *Eur. J. Remote Sens.* **2020**, *53*, 176–191. [CrossRef]
14. Loveland, T.R.; Reed, B.C.; Brown, J.F.; O Ohlen, D.; Zhu, Z.; Yang, L.; Merchant, J.W. Development of a global land cover characteristics database and IGBP DISCover from 1 km AVHRR data. *Int. J. Remote Sens.* **2000**, *21*, 1303–1330. [CrossRef]
15. Gorelick, N. Google Earth Engine. In *EGU General Assembly Conference Abstracts*; American Geophysical Union: Vienna, Austria, 2013; Volume 15, p. 11997.
16. Zhang, C.; Di, L.; Yang, Z.; Lin, L.; Hao, P. AgKit4EE: A toolkit for agricultural land use modeling of the conterminous United States based on Google Earth Engine. *Environ. Model. Softw.* **2020**, *129*, 104694. [CrossRef]
17. Chen, C.; Wang, L.; Myneni, R.B.; Li, D. Attribution of Land-Use/Land-Cover Change Induced Surface Temperature Anomaly: How Accurate Is the First-Order Taylor Series Expansion? *J. Geophys. Res. Biogeosci.* **2020**, *125*. [CrossRef]
18. Myroniuk, V.; Kutia, M.; Sarkissian, A.J.; Bilous, A.; Liu, S. Regional-Scale Forest Mapping over Fragmented Landscapes Using Global Forest Products and Landsat Time Series Classification. *Remote Sens.* **2020**, *12*, 187. [CrossRef]
19. Patel, N.N.; Angiuli, E.; Gamba, P.; Gaughan, A.; Lisini, G.; Stevens, F.R.; Tatem, A.J.; Trianni, G. Multitemporal settlement and population mapping from Landsat using Google Earth Engine. *Int. J. Appl. Earth Obs. Geoinf.* **2015**, *35*, 199–208. [CrossRef]
20. Dong, J.; Xiao, X.; Menarguez, M.A.; Zhang, G.; Qin, Y.; Thau, D.; Biradar, C.; Moore, B. Mapping paddy rice planting area in northeastern Asia with Landsat 8 images, phenology-based algorithm and Google Earth Engine. *Remote Sens. Environ.* **2016**, *185*, 142–154. [CrossRef]
21. Gorelick, N.; Hancher, M.; Dixon, M.; Ilyushchenko, S.; Thau, D.; Moore, R. Google Earth Engine: Planetary-scale geospatial analysis for everyone. *Remote Sens. Environ.* **2017**, *202*, 18–27. [CrossRef]
22. Wang, M.; Zhang, Z.; Hu, T.; Wang, G.; He, G.; Zhang, Z.; Li, H.; Wu, Z.; Liu, X. An Efficient Framework for Producing Landsat-Based Land Surface Temperature Data Using Google Earth Engine. *IEEE J. Sel. Top. Appl. Earth Obs. Remote Sens.* **2020**, *13*, 4689–4701. [CrossRef]
23. Tang, L.; Zhao, M.; Wu, X. Accurate classification of epilepsy seizure types using wavelet packet decomposition and local detrended fluctuation analysis. *Electron. Lett.* **2020**, *56*, 861–863. [CrossRef]
24. Wang, L.; Dong, Q.; Yang, L.; Gao, J.; Liu, J. Crop Classification Based on a Novel Feature Filtering and Enhancement Method. *Remote Sens.* **2019**, *11*, 455. [CrossRef]
25. Xie, Y.; Bie, Q.; Lu, H.; He, L. Spatio-Temporal Changes of Oases in the Hexi Corridor over the Past 30 Years. *Sustainability* **2018**, *10*, 4489. [CrossRef]
26. Venkatappa, M.; Sasaki, N.; Shrestha, R.P.; Tripathi, N.K.; Ma, H.-O. Determination of Vegetation Thresholds for Assessing Land Use and Land Use Changes in Cambodia using the Google Earth Engine Cloud-Computing Platform. *Remote Sens.* **2019**, *11*, 1514. [CrossRef]
27. Brovelli, M.A.; Molinari, M.E.; Hussein, E.; Chen, J.; Li, R. The First Comprehensive Accuracy Assessment of GlobeLand30 at a National Level: Methodology and Results. *Remote Sens.* **2015**, *7*, 4191–4212. [CrossRef]
28. Liu, J.; Peng, S.; Chen, J.; Liao, A.; Zhang, S. Knowledge-based method and engineering practice of globeland30 cultivated land data quality inspection. *Bull. Surv. Mapp.* **2015**, *4*, 42–48. (In Chinese with English Abstract)
29. Martín-Ortega, P.; García-Montero, L.G.; Sibelet, N. Temporal Patterns in Illumination Conditions and Its Effect on Vegetation Indices Using Landsat on Google Earth Engine. *Remote Sens.* **2020**, *12*, 211. [CrossRef]

30. Dadon, A.; Ben-Dor, E.; Karnieli, A. Use of Derivative Calculations and Minimum Noise Fraction Transform for Detecting and Correcting the Spectral Curvature Effect (Smile) in Hyperion Images. *IEEE Trans. Geosci. Remote Sens.* **2010**, *48*, 2603–2612. [CrossRef]
31. Feyisa, G.L.; Meilby, H.; Fensholt, R.; Proud, S.R. Automated Water Extraction Index: A new technique for surface water mapping using Landsat imagery. *Remote Sens. Environ.* **2014**, *140*, 23–35. [CrossRef]
32. Yu, K.-Q.; Zhao, Y.-R.; Liu, Z.-Y.; Li, X.-L.; Liu, F.; He, Y. Application of Visible and Near-Infrared Hyperspectral Imaging for Detection of Defective Features in Loquat. *Food Bioprocess Technol.* **2014**, *7*, 3077–3087. [CrossRef]
33. Chang, C.-I.; Plaza, A. A Fast Iterative Algorithm for Implementation of Pixel Purity Index. *IEEE Geosci. Remote Sens. Lett.* **2006**, *3*, 63–67. [CrossRef]
34. Jensen, J.R. *Introductory Digital Image Processing: A Remote Sensing Perspective (No. Ed. 2)*; Prentice-Hall Inc.: Upper Saddle River, NJ, USA, 1996; p. 379. [CrossRef]
35. Tung, F.; LeDrew, E. The determination of optimal threshold levels for change detection using various accuracy. *Photogramm. Eng. Remote Sens.* **1988**, *54*, 1449–1454.
36. Yang, X.; Pavelsky, T.M.; Allen, G.; Donchyts, G. RivWidthCloud: An Automated Google Earth Engine Algorithm for River Width Extraction from Remotely Sensed Imagery. *IEEE Geosci. Remote Sens. Lett.* **2019**, *17*, 217–221. [CrossRef]
37. Ma, L.; Cheng, L.; Han, W.; Zhong, L.; Li, M. Cultivated land information extraction from high-resolution unmanned aerial vehicle imagery data. *J. Appl. Remote Sens.* **2014**, *8*, 83673. [CrossRef]
38. Yang, C.; Xu, G.; Li, H.; Yang, D.; Huang, H.; Ni, J.; Li, X.; Xiang, X. Measuring the area of cultivated land reclaimed from rural settlements using an unmanned aerial vehicle. *J. Geogr. Sci.* **2019**, *29*, 846–860. [CrossRef]

sustainability

MDPI

Article

Mapping Ecosystem Services in an Andean Water Supply Basin

Diana Marcela Ruíz Ordoñez [1,*,†], Yineth Viviana Camacho De Angulo [1,†], Edgar Leonairo Pencué Fierro [2] and Apolinar Figueroa Casas [1]

1 Grupo de Estudios Ambientales (GEA), Environmental Sciencies, Department of Biology, University of Cauca, Calle 5 #4-70, Popayán 190003, Colombia
2 Optics and Laser Group, Environmental Sciences, Department of Physics, University of Cauca, Calle 5 #4-70, Popayán 190003, Colombia
* Correspondence: dianamruiz@unicauca.edu.co; Tel.: +57-317-3294094
† These authors contributed equally to this work.

Abstract: Socio-ecological dynamics affect the ecosystem services supply and are relevant to generate effective water management strategies; this condition is considered to evaluate under a holistic approach, the water ecosystem services (WES) in an Andean supply basin (ASB) in Colombia. This analysis focus on the connection of biophysical and sociocultural components for the multi-purpose use of water based on The Soil and Water Assessment Tool (SWAT) modelling for Las Piedras River Basin (LPRB). The generated Hydrological Response Units (HRUs), allows to estimate the capacity of the basin for supplying water (quantity) in adequate conditions (quality) for local populations in rural and urban areas, as well as WES zoning. The model was calibrated and validated to generate a baseline scenario, which was complemented with social cartography and participative workshops. The results indicate a low concentration of nitrogen and phosphorus, boosted by specific agro-ecological strategies developed by local communities; however, there are health risks for populations downstream and those that are supplied with water directly from the source. Additionally, Land Use and Land Cover (LULC) affects water availability, which demands restoration and conservation strategies to maintain WES supply for socioeconomic and cultural purposes, since different views on the available WES converge in the basin.

Keywords: ecosystem services supply; planning tool; water pollution; water supply; socioecological conflicts

Citation: Ruíz Ordoñez, D.M.; Camacho De Angulo, Y.V.; Pencué Fierro, E.L.; Figueroa Casas, A. Mapping Ecosystem Services in an Andean Water Supply Basin. *Sustainability* **2023**, *15*, 1793. https://doi.org/10.3390/su15031793

Academic Editor: Yu Cao

Received: 28 November 2022
Revised: 5 January 2023
Accepted: 9 January 2023
Published: 17 January 2023

1. Introduction

The use and ownership of natural resources to meet human needs and reach social wellbeing are concentrated in basins, which guarantee access to water as the enabling component for life, settlements, and economic-productive activities, such as agriculture. However, the interaction between LULC dynamics with climate variability influence the basin's capacity to supply continuous water in adequate quality for urban and rural communities. This transformation is relevant due to the socioecological conflicts that emerge when inequality and inequity for the availability of water (for drinking and household), widen the socio-economic gap, making water management strategies unpredictable and difficult to control.

In the Andean basins, this transformation comes mainly from productive activities that respond to raw materials' demand based on an economic growth model (capitalism), with known environmental liabilities. In the case of Colombia, livestock and agriculture are concentrated in Andean regions (Cauca and Magdalena basins), which in turn support more than 77% of the national population trough water supply and food production [1–4].

In the Upper Cauca River Basin (UCRB) in the Department of Cauca, the agriculture is characterized by small and medium scale crops, that extend toward the páramo (Andean moorland), affecting water sources, native vegetation, and soils, whose effects are intensified due to the uncertainty of climate change and local climate variability [5–8].

Agriculture practices in the UCRB includes deforestation, slash, burn, and the overuse of agrochemical inputs, that affects WES supply, limiting the water availability for local communities [9–11]. These conditions are represented in LPRB, a water supply basin in the southwest of Colombia, which provides multipurpose water for rural communities and drinking water for urban areas where socioecological conflicts around water management are presented.

These affectations to WES supply are related to productive activities, local climate variability and water demands, which have been studied from the biophysical and hydrological valuation approach independently, making analyses to estimate the effects on water supply, runoff dynamics, baseflow, flood events and peak discharges in basins, trough the LULC changes and water quality analysis, but not considering its socioecological integration for management and planning purposes [12,13], as is focused in this research.

Under these approaches, the main tools for analyzing WES are models that simulate the hydrological regulation dynamics based on LULC patterns, for this purpose the SWAT model has been worldwide used for assessing hydrological dynamics [14,15], to analyze freshwater supply and base flow conditions focused on the HRUs as well as the identification and zoning of WES based on hydrological response scenarios [16]. SWAT is also used for studying erosion processes [17], pollution by nutrients [18], basins management strategies [19,20], the monitoring of converting intensive agricultural practices to sustainable practices [21], or in the implementation of payment schemes for environmental services [22].

However, in Colombia, studies that include an integral analysis from a socioecological approach of WES with the SWAT model are not widely used, limiting the opportunities to generate planning tools for local governments.

One of these studies analyzed the climate change, LULC dynamics, and its effects on water yield and carbon sequestration in two Andean watersheds [23]; other research assessed the impacts of changing intensive tillage (IT) for conservation tillage (CT) in a potato crop. A study of the sediment yield, surface runoff and nutrient (nitrogen, phosphorus) losses in surface water runoff [24] was also used to evaluate the water yield in an Andean basin, where the SWAT model was used under different LULC and climate scenarios for water management [19], and in addition to evaluate the impact of LULC on the availability of water resources in conservation areas [25].

Although there are studies for SWAT model implementation at different scales in the country, just a few are developed in the ASB and less in the UPCR, where zoning of WES based on the hydrological dynamics, the relation with productive activities and climate variability are relevant, due to the socioecological dynamics that conditions the water supply for rural communities, increasing their vulnerability.

In this context, this study objective is to produce knowledge of ecosystem services based on hydrological dynamics in ASB, the socioecological conflicts related to WES supply, and the identification of zones for water management, using an integral methodology where participatory workshops and social cartography complements and validate the results of SWAT modeling.

The paper is organized as follows: First, we present an overview of the hydrological model and its application, the dataset and study area, the collection and processing of information. Second, we present the proposed tool for ecosystem services mapping in Andean basins, the results and analysis of applying the tool, and finally, we draw the corresponding conclusions and future developments.

2. Study Area

Las Piedras River basin (LPRB) is located within the municipalities of Popayán and Totoró in the UPCR in southwest of Colombia. It is at 76°31′10″ E and 2°21′45″ N. The basin is composed by two corregimientos (townships): Quintana and Las Piedras, where small peasants and indigenous communities (Puracé and Quintana councils) are located [26].

Figure 1 shows the study area, the river network, the weather stations, and land cover in the basin.

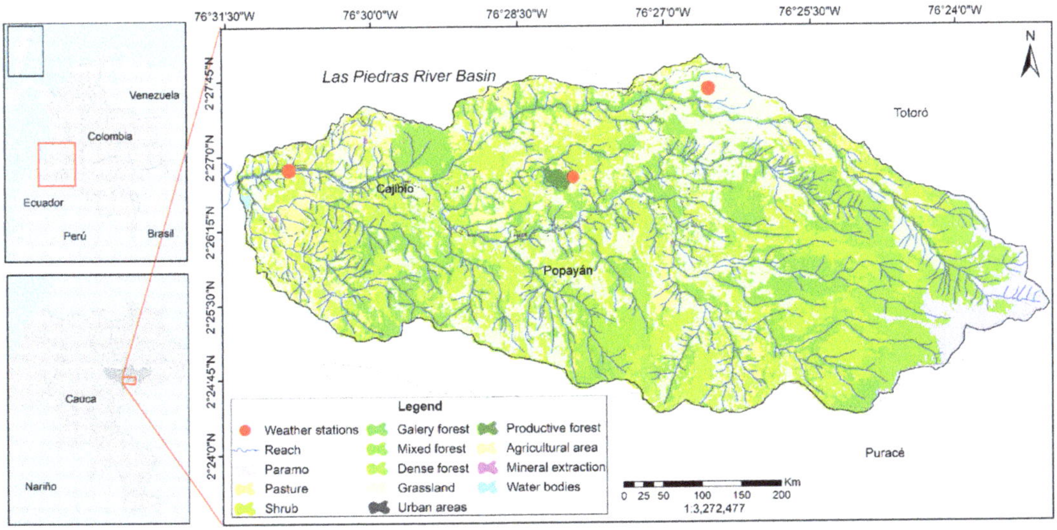

Figure 1. Study area, Las Piedras river basin (LPRB), Cauca, Colombia. In the figure the black framework corresponds to San Andres Island in the Colombian territory and the red framework extend indicator for department of Cauca.

3. Materials and Methods

The study was developed using the Method for Ecosystem Services Mapping (MESM) described in Figure 2. A proposed methodology based on mixed methods research which included (*i*) implementation of the SWAT hydrological model, based on updated cartographic inputs (*ii*) evaluation of ecosystem services (ES) supply complemented with social cartography to locate WES, and the (*iii*) analysis of ES distribution, to understand the socio-ecological conflicts that limit the availability of WES for the LPR communities.

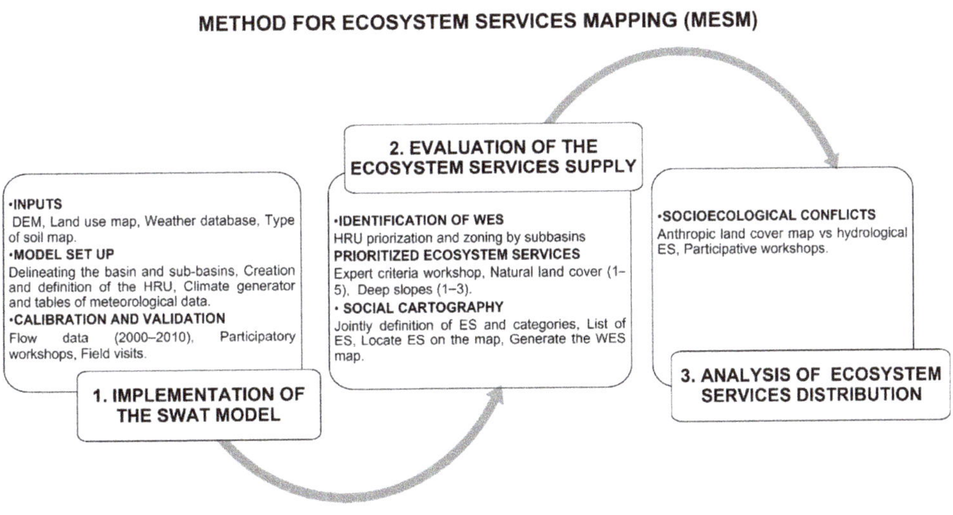

Figure 2. Proposed method for ecosystem services mapping.

3.1. Implementation of the SWAT Model

SWAT is a dynamic and continuous model based on mathematical descriptions of physical, hydro-chemical, and bio-geo-chemical processes that combines elements of physical conditions and vegetation growth processes trough spatial disaggregation or HRUs. This model was developed by the Blackland Research Center in Texas in 1999 for the United States Department of Agriculture (USDA) [27]. SWAT models the basin and its dynamics based on different scenarios, using a semi-distributed deterministic model. It is useful for planning purposes due to the connection of different components of the territory, such as LULC, reforestation activities, population centers and catchment. The model is based on the water balance equation (shown in Equation (1)) to determine the input, output, and storage flows of water in the basin, as well as its hydric response.

$$SW_t = SW_0 + \sum R_{day} - Q_{surf} - E_a - W_{seep} - Q_{gw} \quad (1)$$

where SW_t is the final soil water content (mm); SW_0 is the moisture content in one day i (mm); t is the time (days); R_{day} is the daily precipitation of day i (mm); Q_{surf} is the surface run-off produced of day i (mm); E_a is the evaporation of day i (mm); W_{seep} is the content entering the vadose zone of the soil during day i (mm); Q_{gw} is the flow produced or returned of day i (mm).

The database was created according to the objective of the study and the specific inputs requirements of SWAT model:

3.1.1. Inputs

- Digital elevation model (DEM): for topography, the study used a DEM with 12.5 m accuracy (cell size 12.5 × 12.5), obtained from the Alaska Satellite Facility website; the LPRB has an altitudinal gradient from 1980 to 3820 m.a.s.l.
- Land use map: the map contains information of the areas and landcover types present in LPRB. It was generated for April 2017 (low percentage of clouds) using images of the Sentinel 2A satellite platform, with 10 m precision, considering the Corine Land Cover methodology, adapted in Colombia and the algorithm developed by WP4 RICCLISA [28], identifying 14 landcover types from levels 1, 2, and 3. Field visits and key stakeholders' workshops validated this information (social cartography).
- Weather database: the database was generated from information available of daily precipitation data from nearby weather stations, for the period from 1 January 1999 to 31 December 2017. The statistical weather data required by the SWAT model are the multi-annual averages of maximum and minimum temperature and precipitation, standard deviation for each month, bias coefficient for daily precipitation, number of days of precipitation, probabilities of a humid day after a dry-humid day. These were calculated through the mathematical expressions suggested in the SWAT manual [27].
- Soil type map: contains information of the physical and chemical properties of the LPRB (scale 1:25.000), obtained from information on the study of soils by the planning and management document for LPRB [29].

3.1.2. Model Set Up

- Delineating the basin and sub-basins: The flow direction and the accumulation of water within the sub-basins was simulated with the inputs: DEM, mask of the study area, and the river network, as well as the definition of slope's range and the maximum and minimum elevations. Outlets were selected considering the main drains of the LPRB.
- Creation and definition of the hydrologic response units (HRUs): The HRUs map was based on the superposition of the shapefiles soil types (22 units), land use (12 types), and the specific slopes range (four ranges). From this output, a minimum percentage of aggregation was chosen by expert criteria, considering representative land use, soils, and slopes of the zone, allowing the prioritization of the HRUs, using 1% for LULC, 6% as minimum value for types of soils, and 10% for range of slope, with the

lowest loss of information over an area of the basin and the best distribution in the sub-basins [27].

- Weather generator and tables of meteorological data: Information was included based on the weather station identifiers and location of Arrayanales (ARR) and Diviso (DIV) stations daily precipitation database (mm) and its statistical data. Due to the lack of information in the study area, SWAT model was used to simulate and complete input values of solar radiation, relative humidity, and wind speed.

3.1.3. Calibration and Validation

The calibration of SWAT model for LPRB, was carried out through the SWAT—CUP (SWAT Calibration and Uncertainty Procedures) software with the SUFI2 algorithm [30], which operates through trial and error by randomly changing the values of parameters of interest, such as initial SCS CN II value (Cn2), base flow alpha factor (Alpha_Bf), Groundwater delay (Gw delay), threshold water depth in the shallow aquifer for flow (GWQMN), average slope steepness (slope), saturated hydraulic conductivity (Sol_K), among others. This is done until obtained a reasonable coincidence ($R^2 \geq 0.6$) between the simulation and the values observed.

The model was calibrated and validated with the daily precipitation data (1999–2017) from ARR and DIV weather stations and the monthly streamflow data (1999–2009) from the Puente Carretera (PCA) limnimetric station. Four iterations of 200 simulations each one was carried out, changing the parameters included in the SUFI2. The validation used registries of streamflow (2015–2016), with an iteration of 200 simulations. With the participative workshops and social cartography described in Section 3.2, we contrasted and validated the results from the SWAT model with social perception of the LPRB map.

3.2. Evaluation of the Ecosystem Services Supply

The evaluation of the ES supply was developed under a participative approach with experts and communities, through workshops, social cartography and field visits as established in the proposed MESM, to validate the hydrological modelling, prioritize the HRUs, proposed a joint definition of the concept of ecosystem services, their categories and for zoning each one of them under the land cover/slope combination.

3.2.1. Identification of WES

The WES identification was based on the hydrological conditions of the LPRB modelled with SWAT, with the resulting water-soil-climate-use-slope interaction for the 1999–2017 period, which is represented in the HRUs distribution map.

The SWAT outputs allow the analysis of (*i*) the water production as the water recharge (WYLD) and soil water availability (SW), potential and real evapotranspiration (ET-ETP), and surface runoff (SURQ). Additionally, (*ii*) for the estimation of water pollution, trough the variables of sediments yield and transport (SED YIELD), nitrates on surface runoff (NO3-SURQ) and organic phosphorus (ORGP)

From this, it is feasibly to group SWAT outputs by taking the sub-basin as the unit, to identify the water importance ones, due to the regulating function supported by natural land covers, and those in which is necessary to implement sustainable practices and soil management, considering the productive land covers.

3.2.2. Prioritized Ecosystem Services for LPRB

The HRUs were prioritized with experts and community's stakeholders participative workshops according to (*i*) dominant land cover, assigning importance values for hydrologic regulation from 1 to 5, the most important with one (1) score for natural coverages and the least important in regulation with five (5) score, for anthropic coverages; (*ii*) the slopes with greatest susceptibility to erosion processes [31] were assigned scores corresponding to the value of three (3) for the 0–25% range, a score of two (2) for the 25–75% range, and one (1) for the critical zones (75%).

3.2.3. Social Cartography

Social cartography is a participatory method that combines digital tools with qualitative methods to generate maps that represent the components, relationship and dynamics in specifics landscapes [32], in this case, we conducted workshops with communities of the upper, middle, and lower zones of the LPRB, to carried out four stages: (*i*) The first stage was to produced hand-drawn maps under the community perception to locate the main stream, tributaries, natural forest, crop production areas and pastures for livestock. These maps were then compared and complemented by the stakeholders with a press LULC map, this allowed us to validate the LULC map used as input for the SWAT model; (*ii*) the second stage was to create a jointly definition of ES and the corresponding categories of regulating, cultural and provisioning, then to list ES by each category; (*iii*) the third stage, was the location of each identified ES in the hand-drawn map using words, colors, pictures, or any other symbol to create the legend of the map; and in (*iv*) the fourth stage, the digitalization of this inputs into the HRUs map to generate the WES zoning.

3.3. *Analysis of Ecosystem Services Distribution*

Socioecological Conflicts

The water conflicts for local communities were analyzed trough the stakeholders (institutions and communities) perspectives of the water supply dynamics in LPRB, this was carried within a workshop to discuss about the different uses and views that each one of the stakeholders have respect to the LPRB as well as the action or strategy developed for its water management, this discussion was based on the results of SWAT model, the LUCL dynamics and WES zoning. The guiding questions for conducting the discussion about the use was: What do you use water for on your farm/home? For analyzing the views we ask the question: What does the LPRB represent for you, your family and/or community? For actions implemented the question was: What individual or community actions are developed to conserve water? Each question was discussed in focus groups of institutional, small peasants' and indigenous communities' representatives; the final output was a single statement that represents the collective thought.

4. Results

4.1. *Implementation of the SWAT Model*

This section presents the updated cartographic outputs for LPRB, as a result of the SWAT modelling, such as the total area of the basin, the tributaries, the calibration parameters, and the hydrograph.

4.2. *Updated Outputs of the SWAT Modelling of LPRB*

The LPRB delimitation was updated with a total area of 6606.27 ha, approximately 20 ha less than the one reported by [26], due to the precision of the DEM used in the study, from 30 m to 12.5 m. Based on this, 18 sub-basins were identified in the LPRB, compared to 13 reported by [33], regarding to this, the Robles sub-basin is included (6), the Santa Teresa sub-basin is divided into Santa Teresa (2), Las Pavas (3) and Santa Teresa II (4), the Aguas Claras sub-basin is divided into Aguas Claras (13), La Cabaña (14) and San Pedro (15) And the Buena Vista sub-basin is divided into the El Cedro (17), Peñas Blancas (18) and Piedra Negra (16), as shown in Figure 3 for details and comparisons.

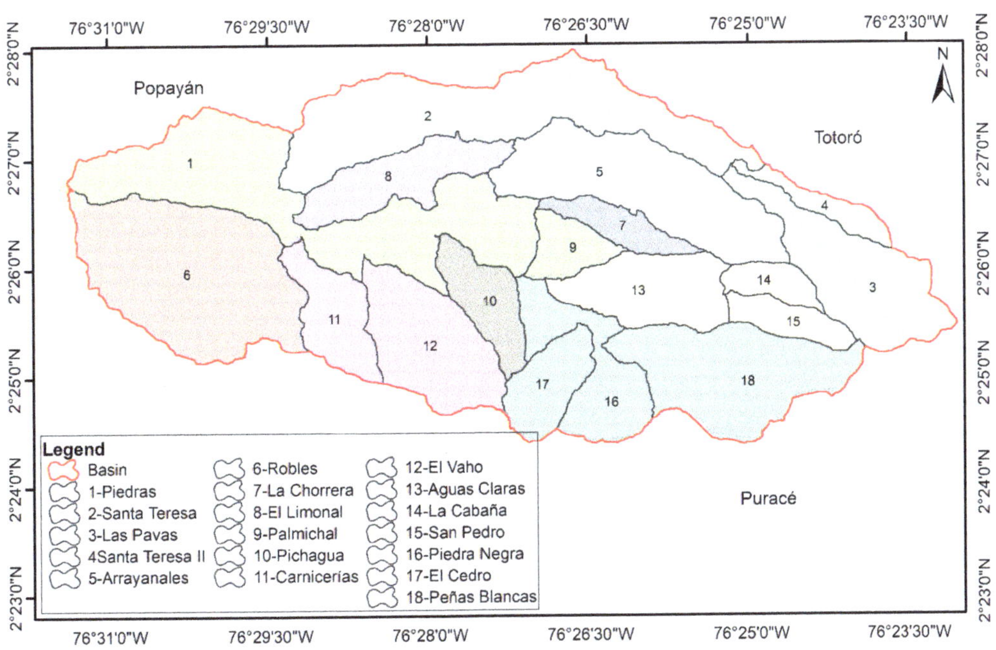

Figure 3. Sub-basins obtained with SWAT (delineated and numbered), compared with the sub-basins reported by [33] (differentiated by colors).

4.3. SWAT Calibration and Validation

The values of the determination coefficient were $R^2 = 0.99$ for ARR and $R^2 = 1$ for DIV. Water quality data were not considered in the calibration because these do not complement the minimum historical data, however, satisfactory calibration was obtained with $R^2 = 0.614$ [34] for monthly streamflow data (Figure 4), Table 1 shows the calibrated values for the parameters of interest.

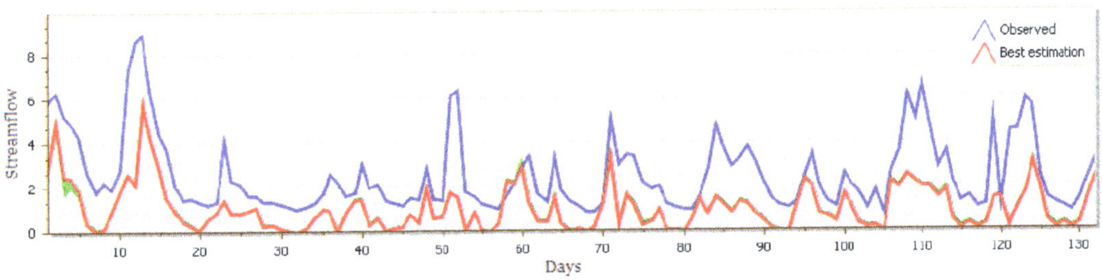

Figure 4. Simulated and observed streamflow hydrograph.

Table 1. Calibration parameters of the SWAT model for LPRB.

ID	Parameter	Description	Process	Initial Range	Calibrated Value
1	GWQMN (Threshold water depth in the shallow aquifer for flow)	Threshold of water depth	Base flow	550–1000	862.96
2	Alpha-Bf (Base flow alpha factor)	Base flow factor	Base flow	0–1	0.5
3	Gw-Delay (Groundwater delay)	Storage of groundwater	Base flow	0–50	26.86
4	Cn2 (Initial SCS CN II value)		Run-off	35–98	45.63

4.4. Evaluation of the ES Supply

This section presents the identification of WES as result of the hydrological modelling, which stablishes the baseline conditions for water supply in LPRB, the identification and the prioritization of WES according to the stakeholders' perspectives and the zoning of WES under the communities' views.

4.5. Identification of WES

A 607 HRUs map of the LPRB was prioritized (from an initial 1687 HRU map) by a minimum percentage of aggregation considering representative land use, soils, and slopes of the zone, covering 100% of the basin modeled with the best distribution in the 18 sub-basins [27], from this output. The actual condition of the LPRB was modelled made at sub-basins level; the results are shown in Figure 5.

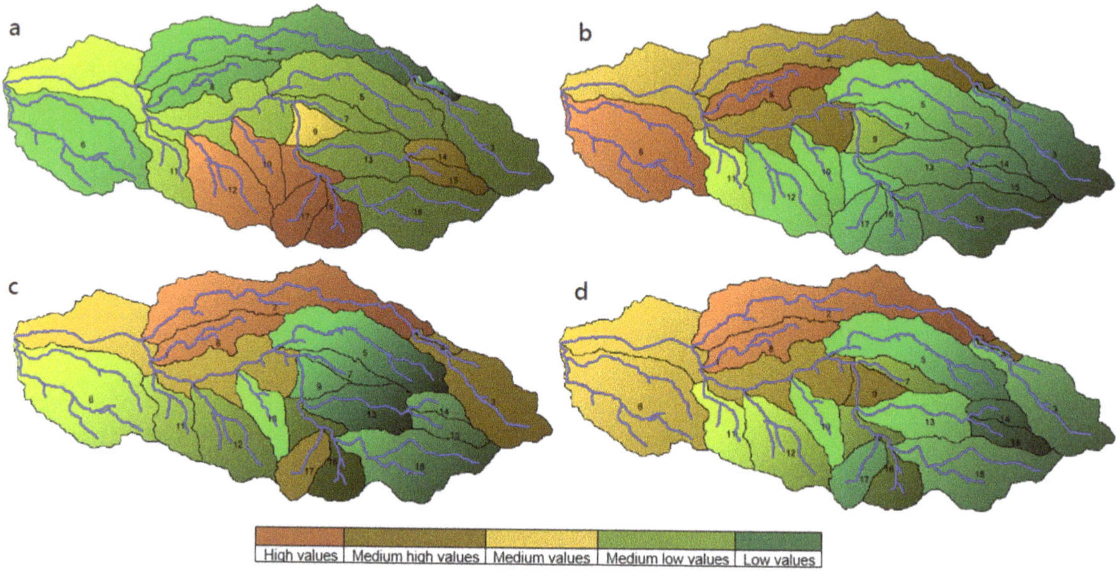

Figure 5. Hydrological simulation of LPRB. (**a**) Water recharge (WYLD). (**b**) Amount of available soil water (SW). (**c**) Sediment yield (SED YIELD). (**d**) Nitrates on surface run-off (NO3-SURQ).

The water production in the LPRB, i.e., is represented by the amount of water generated in each sub-basin reaching the streamflow, Figure 5a,b, shows the offer of WES by the parameters WYLD and SW, corresponding to water balance in the land phase [35,36]. Thus, it was possible to identify the sub-basins with the highest water contribution (WYLD), corresponding to areas with prevalence of high slopes (≥75%) and natural land cover (dense forest, páramo, and shrub (Figure 1) such as the sub-basins 17 (856.92 mm), 3 (664.13 mm), 10 (818.56 mm), 16 (808.83 mm), 12 (798.89 mm), and 14 (740.54 mm).

The green color identifies the sub-basins with the lowest water recharge, which are areas with crops and grasslands (clean and degraded), in the 4, 2, 8, 6, and 1 sub-basins. The amount of water stored in the soil (SW) for plants, increases toward lower zones, especially in sub-basin 2 (748.13 mm), 8 (803.49 mm) and 6 (872.06 mm), corresponding to productive areas. The water losses from soil surface in the LPRB, were analyzed under weather conditions simulated with the Penman–Monteith method (1999–2017), for the ET and ETP. Natural covers of the upper area had high values of ETP in sub-basin 3 (3500 mm), 15 (3497 mm) and 18 (3473 mm) while crops and pastures of the lower area presented high values of ET in sub-basins 2 (947.69 mm), 8 (942.95 mm), 6 (936.52 mm) and 9 (827.63 mm). The water consumption by forests is greater than in other vegetation types due to the depth of roots, height, and foliage. Water is retained and stored in the soil-vegetation interphase and regulates source recharge processes, while zones with higher ET are susceptible to drought due to poor retention and regulation of soil water [37]. In these dynamics, weather conditions, productive and management practices are determinant for hydrological regulation, in the case of the LPRB, the areas with dominant agricultural and fish farming activities are sub-basins with high ET values.

With respect to water quality, in the LPRB, the SED YIELD parameter shown in Figure 5c resembles the nutrient loss response, where the sub-basins 2 (0.88 ton/ha), 8 (0.913 ton/ha) and 1 (0.49 ton/ha) presented higher accumulation of contaminants in the soil (in brown) from agro-chemicals of the potato's crops.

According to the NO3-SURQ parameter, the largest yields were presented in sub-basins 8 (1.11 kg/ha), 2 (1.06 kg/ha), 4 (0.98 kg/ha) and 1 (0.81 kg/ha), related to the use of agro-chemicals for crops and livestock production Figure 5d. The sub-basins 1 (0.81 kg/ha), 9 (0.79 kg/ha), 13 (0.44 kg/ha), and 14 (0.32 kg/ha), had lower crop production because these are conservation areas with fragile ecosystems and low-fertility soils. The results showed a similar behavior for phosphorous and nitrogen, the largest producers of ORGP were the 1 (1.57 kg/ha), 2 (3.41 kg/ha), 8 (3.94 kg/ha) and 11 (1.40 kg/ha) sub-basins, related to the larger monocrop and livestock production zones. The distribution of this nutrient is key in management processes because it comes from both organic (ash, manure) and chemical (commercial) sources, and is enhanced in scenarios of excessive fertilization combined with soil compaction caused by cattle trampling.

4.6. Prioritized Ecosystem Services for LPRB

The joint definition of ES established by local communities of the LPRB is as follows: *"Ecosystem services are what nature provides to people, it results from interaction with human beings, where man receives benefits"*. To understand the specific categories of ES, communities in LPR established that regulating ES represents: *"Equilibrium in the biological processes of ecosystems"*; the cultural ES are: *"Goods and materials that contribute to inner wealth"*; and finally, the provisioning ES refer to: *"What nature gives us"*. The specific WES identified and classified by the upper, middle, and lower zone of the LPR, are presented in Table 2.

Table 2. Prioritized ecosystem services for LPRB.

ES CATEGORY	PRIORITIZED ES Upper	 Middle	 Lower
Provisioning	▪ Food sovereignty ▪ Water availability for the communities	▪ Water quantity ▪ Water quality	▪ Productivity availability of timber resources ▪ Good quality water
Regulating	▪ Air regulation ▪ Climate regulation ▪ Hydrological regulation	▪ Oxygen availability ▪ Biological control ▪ Climate regulation	▪ Soil nutrient cycling ▪ Pollination ▪ Oxygen availability
Cultural	▪ Sacred sites ▪ Maintenance of oral tradition ▪ Knowledge of the territory	▪ Traditional knowledge ▪ Maintenance of oral tradition.	▪ Field schools ▪ Ecotourism areas ▪ Traditional knowledge of the territory ▪ Network of civil society reserves for conservation

4.7. Social Cartography

According to the WES map (Figure 6) the upper zone of the LPRB was related mainly with regulation and cultural ES, associated with zones of natural regulating coverages that are identified as sacred or pilgrimage sites like the Puzná Mountain. The middle zone had an important supply of cultural ES related with property appraisals and the ecotourism potential, because of their strategic location toward zones with high slopes, conservation, and areas of environmental protection. The lower zone represented the availability of provisioning ES, associated in this case with grazing and crop areas.

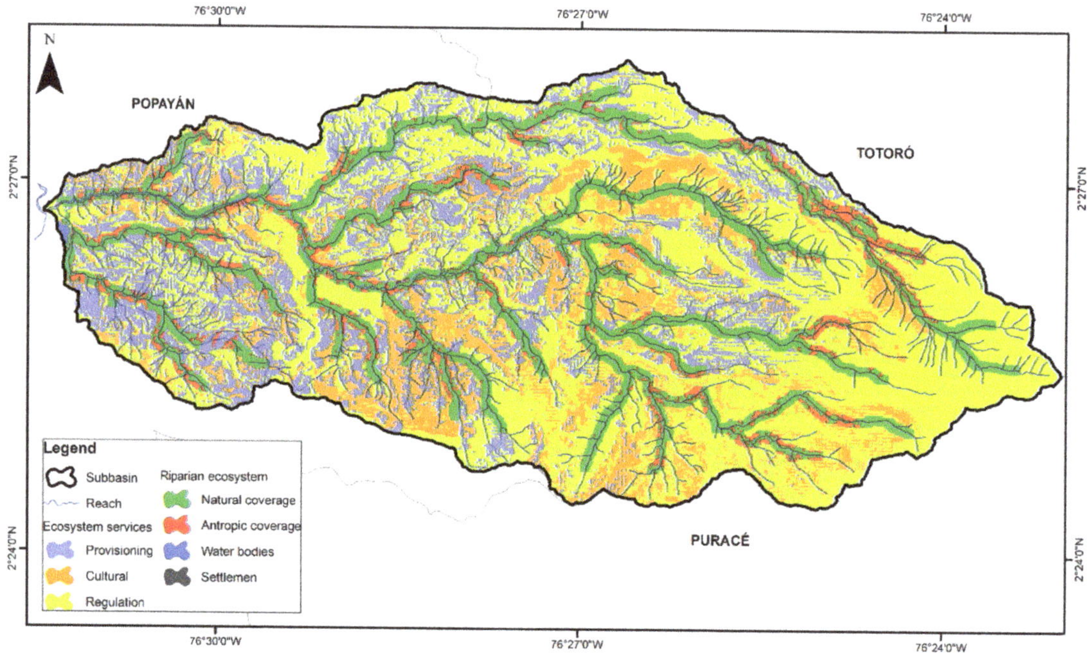

Figure 6. Distribution of WES in the LPRB.

The ES analysis evidenced the community relationships with their territory trough a deep connection with the LPRB "*the water connects us*", where the provider and beneficiary stakeholders' dynamics, as well as the productive activities developed, condition the opportunities to sustainable socioecological transition processes.

The ES approach has been incorporated into the management strategies of some local entities, such as the ES payments in farmers with water recharge areas, which includes a property tax discount by the municipal administration (Agreement 30/2012) and environmental educational processes. These sorts of payments are applicable to rural properties located in areas with hydrological importance for the water intake of the municipal water service, that has been recognized by the environmental authority. This management strategy is included in the planning and environmental conservation processes of the municipal aqueduct enterprise. From local communities' leadership, an important strategy for conservation is the creation of civil society's nature reserves (a formal protection figure recognized by Colombian environmental ministry) and the natural reserves that still are not recognized by local government, which in turn, constitutes important places for ecotourism routes.

4.8. *Sociecological Conflicts*

The production areas established in the riparian buffer zone (Figure 6), represent an important source of trade-offs between ES, such as food supply, the regulation of water quality, barriers of sediments, and nutrients from hillside areas, where livestock has established. Additionally, these crops have strong exposure to the effects of prolonged periods of rainfall, LPRB communities indicated that drought effect is exacerbate by the soils' low fertility and steep slopes, causing socioeconomic affectation to the families, that depends entirely on the agriculture and livestock.

These productive activities on the riparian and hillside area were relevant in the sub-basins of the upper zone of the LPRB, such as sub-basins 2, 5, and 8, with extended potato, bean, and corn crop areas, which are produced under conventional schemes with agrochemical inputs, logging, and burning practices. Although the main production activity in the LPRB is livestock, conventional agriculture practices and fish farming are important sources of pollution, as the SWAT model reveals with respect to the areas with this type of activities. Thus, these production processes limit the supply of WES, due to affectations on water quality and quantity.

From the stakeholder's analysis, it was possible to identify the key points of divergences and convergences with respect to their views of the LPRB, as well as the management actions taken by each one of them under these views. That is, the institutional actors consider the LPRB as a water production space for drinking water; for small peasants, it represents the territory for productive activities and multipurpose water supply that supports their socioeconomic needs, and for indigenous communities, it is the space for silvopastoral systems and multipurpose water supply, where community coexist with nature in an ancient right where the territorial expansion is needed. Due to these different views of the LPRB, and despite the peace and convivence agreement signed between local communities, it has been difficult to articulate the water management strategies; some of them are presented in Figure 7.

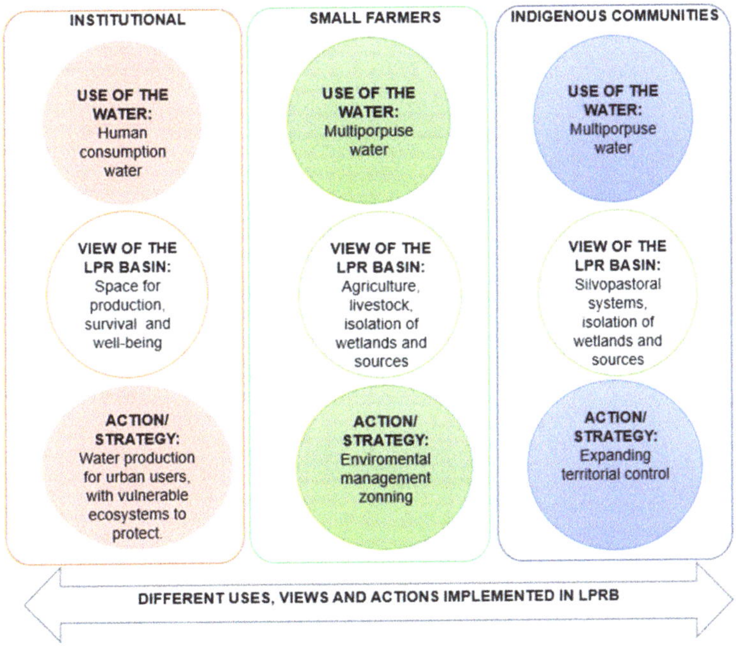

Figure 7. Different stakeholders' perspectives of the water supply in LPR.

5. Discussion

One of the most important contributions of this study is the possibility of performing WES supply analysis in watersheds of water importance, such as those of the Andean zone, which do not have detailed or historical inputs for the implementation of hydrological models. The MESM offers a methodological integration that draws on the strengths of quantitative and qualitative methods.

5.1. Implementation of the SWAT Model

The resolution of the DEM (12.5 m) improves the landscape shapes, making the delimitation of the LPRB contour more precise with an area approximately 20 ha less than reported in previous studies [26,33,38], as well as the identification of the 18 sub-basins, six

(6) more than reported by [38] and seven (7) more than reported by [33]; additionally, we confirm that sub-basin 6 is included in the LPRB area, as reported by [33].

For calibration, the monthly values streamflow observed and the best estimated shown in Figure 4; corresponds to January 1999 and December 2010 (R^2 = 0.614), The most distant values are because the simulation was made without taking into account the macro-scale weather conditions, as the Southern Oscillation (ENSO) with the warming phase or Niño and the cold phase or Niña, due to the missing historical data; in the region the ENSO was presented during five years (1999, 2000, 2007, 2008, 2011). This caused an underestimation in the simulated streamflow, so it is relevant to mention that the availability of climate data with a minimum historical record of 15 years is important to improve the calibration of the model and to compensate for the lack of climate information [34]. The SWAT model validation was complemented with social cartography and participatory workshops; this allows to improve the LULC map, one of the most important inputs for SWAT, as well as providing a space for knowledge dialogue with stakeholders.

Through the HRUs map, is possible to identify that the largest extension of HRUs corresponds to anthropized cover (grassland) related to livestock, the main productive activity in the LPRB [33]. The natural regulating land covers, such as dense forest, is low (11.55% of the sub-basin area), located in the higher zones, where agricultural and livestock is limited by the conditions of the terrain with pronounced slopes and hillside areas, this corresponds to conservation areas isolated by stakeholders The natural pasture cover of the entire LPRB, is an area of socioeconomic interest because it is a potential area for expanding urban and productive activities.

5.2. Evaluation of the ES Supply

The WES related to water quantity, represented by the parameters WYLD and SW, indicate that the sub-basins with the higher water supply (WYLD) are located towards the upper areas, where there is less presence of productive activities, the sub-basins with the greatest amount of water stored in the soil (SW) are located in areas with soils with moderate agrological capacity, where medium-scale productive practices are developed. The ET is higher in the sub-basins of the middle and lower zones, due to the cultivated areas, while in the upper zone, the ETP is higher due to the evaporation processes of rainfall intercepted by the canopy and tree transpiration of the paramo [37].

The sub-basins that contribute the greatest concentrations of nutrients and sediments are Limonal (8), Santa Teresa (2), Santa Teresa II (4), and Las Piedras (1); these areas have productive activities, but soils have moderate to low fertility and agrological capacity, which demands the implementation of sustainable production systems, especially in upper sub-basins 2 and 4.

To estimate the dynamics of the WES related to quality, we analyze the N and P cycle associated with production practices in the sub-basin which are low, since the use of agrochemicals is not generalized and there are transitions towards the use of organic agro-inputs; however, this condition is not common to the supply sources of the department of Cauca or the Andean zone [10,17]. Although the values calculated are low with respect to the whole basin area, these processes are directly related with the water quality changes of the source basin for the municipal water service, affecting the potabilization processes required for urban users, increasing fees, continuity, and quality of the water supply [19,23].

In this sense, the supply of WES is related to the needs of the communities in the availability of water and food for the LPRB inhabitants and for human consumption in urban areas, but in both cases, there are health risks for populations, on the one hand by drinking water directly from the source and on the other hand by the potabilization requirements. From the socioecological view of the LPRB, we identify problems with access to water for LPRB inhabitants, by quantity (high and medium zones) and quality (low zone) and although the agricultural production is in the process of converting to sustainable practices, the fertility limitations that characterize the LPRB soils must be overcome, and the commercialization channels must be improved.

5.3. Analysis of ES Distribution

Community groups of the LPRB evidence their differences through the prioritization of cultural ES; in the upper zone (mainly indigenous communities) we identify ES related to knowledge of the territory and the collective cultural heritage. In the lower zone, organizational and associative activities of the small farmers communities prevailed, and in the middle zone, we find a transition between the two indigenous-small peasants' visions, by prioritizing aspects that give greater "value" to their lands in the sense of environmental importance with ecotourism areas [13].

Because of this, it is necessary to strengthen the social network through the articulation of community and institutional stakeholders. To generate synergies for conducting planning and management processes, such actions must be accompanied by strategies for improving socioeconomic conditions of the local inhabitants, where their broad knowledge and environmental sense could be included in community-based productive alternatives that allow WES supply and better-quality life.

6. Conclusions

In this paper, we present an integrated analysis of the supply of water ecosystem services in a strategic Andean water supply basin in Colombia. The results show the high susceptibility to hydric erosion due to changes in texture and structure of the soils in the LPRB, which is the result of continuous implementation of agricultural activities with inadequate technologies. This condition affects the availability of nutrients, generates loss of soil fertility, and increases run-off rates, related, in turn, with dynamics in nutrient concentration and alteration of the pH in water from the stream. Additionally, the middle and lower zones of the LPRB are in drought risk due to the poor retention and regulation of water in the soil, as indicated by high ET values associated with crop areas.

According to the SWAT model, we identify the sub-basins that demand restoration and conservation actions, because of their hydrological importance: Las Pavas (3), Pichagua (10), and San Pedro (15), as well as the sub-basins where it is necessary to strengthen the processes of soil management as they represent areas with a predominant anthropic land cover, Santa Teresa (2), Limonal (8), and Cedro (17), and productive activities for sustain local communities. Additionally, the sediment production and transport in the LPRB is higher in the Santa Teresa (2), Limonal (8), and Piedras (1) sub-basins, related to agrochemicals, used for potatoes' crops, while the Arrayanales (5) and La Chorrera (7) sub-basins show low sediment accumulation, because these are conservation areas delimited by stakeholders.

As future developments, we would like to consider the modelling of management scenarios with: (i) diffuse pollution processes and soil compaction, in relation to the main productive activities developed in the basin that are affecting hydrological ecosystems services; and (ii) the interactions between inhabitants of the LPRB in the rural area and uses of the water supply system in the urban area.

Author Contributions: Conceptualization, methodology, validation, formal analysis, writing—review and editing D.M.R.O. and Y.V.C.D.A.; resources, supervision, review and editing E.L.P.F.; writing—review and editing A.F.C. All authors have read and agreed to the published version of the manuscript.

Funding: The authors would like to thank the different entities involved in providing the funding to develop this research work. This work was supported by Universidad del Cauca (501100005682) under Grant number (BPIN 2020000100714) ID 5650 and ID 5142 supported by the Water Security and Sustainable Development Hub funded by the UK Research and Innovation's Global Challenges Research Fund (GCRF) [grant number: ES/S008179/1]. this work was performed with the support of the Department of Science Technology and Innovation Colciencias in Colombia, through the bicentennial scholarship program "Young research and doctoral students—567/2012".

Institutional Review Board Statement: The study was conducted in accordance with the Declaration of Helsinki, and approved by the Ethics Committee of the University of Cauca (Protocol 6.1–1.25/23, 2 December 2020).

Informed Consent Statement: Informed consent was obtained from all subjects involved in the study.

Data Availability Statement: Not applicable.

Acknowledgments: This research was performed with support from the Ministry of Science Technology and Innovation, MINCIENCIAS, in Colombia, through the Bicentennial Scholarship Program "Young researchers and doctoral students", the Water Security and Sustainable Development Hub funded by the UK Research and Innovation's Global Challenges Research Fund (GCRF) (grant number: ES/S008179/1) and the project "Vulnerability and Risk in Drinking Water Systems in Cauca—AQUARISC" financed by the science ministry in Colombia. Gratitude is expressed to the Las Piedras River Foundation, the Environmental Studies and Optical and Laser Research Groups at Universidad del Cauca. The authors also thank the indigenous and small farming communities living in the LPR basins and Doctor Tatiana Solano for her review and contributions.

Conflicts of Interest: The authors declare no conflict of interest.

References

1. González, R.B. Apertura y reprimarización de la economía colombiana. *Nueva Soc.* **2011**, *231*, 46–65.
2. Vergara, W.V. Reforma agraria en Colombia y 'Prosperidad para Todos': El camino hacia el desarrollo humano. *Rev. Univ. Salle* **2011**, *54*, 87–109.
3. IDEAM. *Estudio Nacional del Agua*; Ministro de Ambiente y Desarrollo Sostenible: Bogota, Colombia, 2018.
4. Berrouet, L.; Villegas-Palacio, C.; Botero, V. Vulnerability of Rural Communities to Change in an Ecosystem Service Provision: Surface water supply. A Case Study in the Northern Andes, Colombia. *Land Use Policy* **2020**, *97*, 104737. [CrossRef]
5. Ruiz, D.; Moreno, H.A.; Gutiérrez, M.E.; Zapata, P.A. Changing climate and endangered high mountain ecosystems in Colombia. *Sci. Total Environ.* **2008**, *398*, 122–132. Available online: http://www.sciencedirect.com/science/article/B6V78-4SBRTJ2-5/2/1dfd957a2f5d7a1c6b4a7cdbec927661 (accessed on 15 September 2018). [CrossRef]
6. van der Hammen, T. Diagnóstico, cambio global y conservación. In *Congreso Mundial de Páramos*; Ministerio de Medio Ambiente: Paipa, Colombia, 2002; Volume 1, pp. 60–71.
7. Buytaert, W.; Iñiguez, V.; de Bièvre, B. The effects of afforestation and cultivation on water yield in the Andean páramo. *Ecol. Manag.* **2007**, *251*, 22–30. Available online: http://www.sciencedirect.com/science/article/B6T6X-4P9K8T4-2/2/6325e25837319cc9affcb748d21a719a (accessed on 1 April 2019). [CrossRef]
8. Harden, C. Human impacts on headwater fluvial systems in the northern and central Andes. *Geomorphology* **2006**, *79*, 249–263. [CrossRef]
9. Buytaert, W.; Cuesta, F.; Tobón, C. Potential impacts of climate change on the environmental services of humid tropical alpine regions. *Glob. Ecol. Biogeogr.* **2011**, *20*, 19–33. [CrossRef]
10. Crespo, P.; Célleri, R.; Buytaert, W.; Feyen, J.; Iñiguez, V.; Borja, P.; De Bièvre, B. Land use change impacts on the hydrology of wet Andean páramo ecosystems. *Status Perspect. Hydrol. Small Basins* **2010**, *336*, 71–76.
11. Shen, Z.; Qiu, J.; Hong, Q.; Chen, L. Simulation of spatial and temporal distributions of non-point source pollution load in the Three Gorges Reservoir Region. *Sci. Total Environ.* **2014**, *493*, 138–146. [CrossRef]
12. Francesconi, W.; Srinivasan, R.; Pérez-Miñana, E.; Willcock, S.P.; Quintero, M. Using the Soil and Water Assessment Tool (SWAT) to model ecosystem services: A systematic review. *J. Hydrol.* **2016**, *535*, 625–636. [CrossRef]
13. Rincón-Ruiz, A.; Arias-Arévalo, P.; Hernández, J.M.N.; Cotler, H.; Caso, M.A.; Meli, P.; Tauro, A.; Akerberg, V.D.Á.; Avila-Foucat, V.S.; Cardenas, J.P.; et al. Applying integrated valuation of ecosystem services in Latin America: Insights from 21 case studies. *Ecosyst. Serv.* **2019**, *36*, 100901. [CrossRef]
14. Hoyos, N.; Correa-Metrio, A.; Jepsen, S.M.; Wemple, B.; Valencia, S.; Marsik, M.; Doria, R.; Escobar, J.; Restrepo, J.C.; Velez, M.I. Modeling Streamflow Response to Persistent Drought in a Coastal Tropical Mountainous Watershed, Sierra Nevada De Santa Marta, Colombia. *Water* **2019**, *11*, 94. [CrossRef]
15. Lehmann, A.; Timoner, P.; Fasel, M.; Lacayo, M.; Vaghefi, S.A.; Abbaspour, K.C. SWATCH21: A project for linking eco-hydrologic processes and services to aquatic biodiversity at river and catchment levels. *Ecohydrol. Hydrobiol.* **2019**, *19*, 182–197. [CrossRef]
16. Glavan, M.; Pintar, M.; Urbanc, J. Spatial variation of crop rotations and their impacts on provisioning ecosystem services on the river Drava alluvial plain. *Sustain. Water Qual. Ecol.* **2015**, *5*, 31–48. [CrossRef]
17. Uribe, N.; Srinivasan, R.; Corzo, G.; Arango, D.; Solomatine, D. Spatio-temporal critical source area patterns of runoff pollution from agricultural practices in the Colombian Andes. *Ecol. Eng.* **2020**, *149*, 105810. [CrossRef]
18. Rajaei, F.; Dahmardeh Behrooz, R.; Ahmadisharaf, E.; Galalizadeh, S.; Dudic, B.; Spalevic, V.; Novicevic, R. Application of Integrated Watershed Management Measures to Minimize the Land Use Change Impacts. *Water* **2021**, *13*, 2039. [CrossRef]
19. Villamizar, S.R.; Pineda, S.M.; Carrillo, G.A. The Effects of Land Use and Climate Change on the Water Yield of a Watershed in Colombia. *Water* **2019**, *11*, 285. [CrossRef]

20. Mera-Parra, C.; Oñate-Valdivieso, F.; Massa-Sánchez, P.; Ochoa-Cueva, P. Establishment of the Baseline for the IWRM in the Ecuadorian Andean Basins: Land Use Change, Water Recharge, Meteorological Forecast and Hydrological Modeling. *Land* **2021**, *10*, 513. [CrossRef]
21. Wei, X.; Garcia-Chevesich, P.; Alejo, F.; García, V.; Martínez, G.; Daneshvar, F.; Bowling, L.C.; Gonzáles, E.; Krahenbuhl, R.; McCray, J.E. Hydrologic Analysis of an Intensively Irrigated Area in Southern Peru Using a Crop-Field Scale Framework. *Water* **2021**, *13*, 318. [CrossRef]
22. Bennett, D.E.; Gosnell, H. Integrating multiple perspectives on payments for ecosystem services through a social–ecological systems framework. *Ecol. Econ.* **2015**, *116*, 172–181. [CrossRef]
23. Clerici, N.; Cote, N.F.; Escobedo, F.J.; Rubiano, K.; Villegas, J.C. Spatio-temporal and cumulative effects of land use-land cover and climate change on two ecosystem services in the Colombian Andes. *Sci. Total Environ.* **2019**, *685*, 1181–1192. [CrossRef] [PubMed]
24. Uribe, N.; Corzo, G.; Quintero, M.; van Griensven, A.; Solomatine, D. Impact of conservation tillage on nitrogen and phosphorus runoff losses in a potato crop system in Fuquene watershed, Colombia. *Agric. Water Manag.* **2018**, *209*, 62–72. [CrossRef]
25. Ortegón, Y.A.C.; Acosta-Prado, J.C.; Castellanos, P.M.A. Impact of Land Cover Changes on the Availability of Water Resources in the Regional Natural Park Serranía de Las Quinchas. *Sustainability* **2022**, *14*, 3237. [CrossRef]
26. Ruiz, D.M.; Martínez, J.P.; Otero, J.D.; Figueroa, A. Effects of productive activities on the water quality for human consumption in an Andean basin, a case study. *Rev. Int. Contam. Ambient.* **2017**, *33*, 361–375. [CrossRef]
27. Arnold, J.G.; Moriasi, D.N.; Gassman, P.W.; Abbaspour, K.C.; White, M.J.; Srinivasan, R.; Santhi, C.; Harmel, R.D.; Van Griensven, A.; Van Liew, M.W.; et al. SWAT: Model use, calibration, and validation. *Trans. ASABE* **2012**, *55*, 1491–1508. [CrossRef]
28. Pencué, E.L.; Solano, Y.T.; Corrales, J.C.; Figueroa, A. A Semi-Supervised Hybrid Approach for Multitemporal Multi-Region Multisensor Landsat Data Classification. *IEEE J. Sel. Top. Appl. Earth Obs. Remote Sens.* **2016**, *9*, 5424–5435. [CrossRef]
29. CRC. *Plan de Ordenación y Manejo de la Subcuenca Hidrográfica del Río Las Piedras (Actualización) Resolucion 0751/2006. F. P. R. L. Piedras. Popayán, CRC: 457*; CRC: Boca Raton, FL, USA, 2006.
30. Kumar, N.; Singh, S.K.; Srivastava, P.K.; Narsimlu, B. SWAT Model calibration and uncertainty analysis for streamflow prediction of the Tons River Basin, India, using Sequential Uncertainty Fitting (SUFI-2) algorithm. *Model. Earth Syst. Environ.* **2017**, *3*, 30. [CrossRef]
31. IGAC. *Estudio General de Suelos y Zonificación de Tierras Departamento del Cauca*; Instituto Geografico Agustin Codazzi: Bogota, Colombia, 2009; Volume 2, p. 86.
32. Ruíz, D.; Maysels, R.; Muñoz, F.F.; Díaz, E. Water Security Hub. In *Social Cartography Participatory Workshops*; GCRF-UKRI: Newcastle, UK; Available online: https://www.watersecurityhub.org/rmgc/social-cartography-participatory-workshops-qualitative-digital-methods-module (accessed on 27 November 2022).
33. Mejia, L.R. Manejo Adaptativo del Territorio en una Cuenca Altoandina Desde la Diversidad Cultural y Ecosistémica. Ph.D. Thesis, de Doctorado en Ciencias Ambientales. Universidad del Cauca, Popayán, Colombia, 2017.
34. Santhi, C.; Arnold, J.G.; Williams, J.R.; Dugas, W.A.; Srinivasan, R.; Hauck, L.M. Vadidation of the SWAT model on a large RWER basin with point and nonpoint sources 1. *JAWRA J. Am. Water Resour. Assoc.* **2001**, *37*, 1169–1188. [CrossRef]
35. Jin, G.; Deng, X.; Hasan, S.S.; Zhao, C.; Gibson, J. Hydrological Ecosystem Services for Integrated Water Resources Management. *Ecohydrology* **2018**, 361–386. [CrossRef]
36. Kenward, R.E.; Whittingham, M.J.; Arampatzis, S.; Manos, B.D.; Hahn, T.; Terry, A.; Simoncini, R.; Alcorn, J.; Bastian, O.; Donlan, M.; et al. Identifying governance strategies that effectively support ecosystem services, resource sustainability, and biodiversity. *Proc. Natl. Acad. Sci. USA* **2011**, *108*, 5308–5312. [CrossRef]
37. Caro-Camargo, C.A.; Velandia-Tarazona, J.E. The effect of changes in vegetation cover on the hydrological response of the sub-basin Los Pozos. *Dyna* **2019**, *86*, 182–191. [CrossRef]
38. Campaz, G.; Enríquez, J. *Simulación de Caudales en la Subcuenca Río Las Piedras Mediante la Aplicación del Modelo Hidrológico SWAT*; Universidad del Cauca: Popayán, Colombia, 2012.

sustainability

MDPI

Article

Construction and Optimization of an Ecological Network in the Comprehensive Land Consolidation Project of a Small Rural Town in Southeast China

Mengyuan Su [1], Xiaoqian Fang [1], Kaiying Sun [1], Jiahao Bao [1] and Yu Cao [1,2,*]

[1] Department of Land Management, School of Public Affairs, Zhejiang University, Hangzhou 310058, China
[2] Land Academy for National Development, Zhejiang University, Hangzhou 310058, China
* Correspondence: caoyu@zju.edu.cn

Abstract: In recent years, China has put forward comprehensive land consolidation projects to solve problems in rural areas, such as cultivated land fragmentation, scattered spatial pattern of construction land and ecological environment pollution, and boost the rural revitalization strategy. Constructing ecological networks is important for maintaining ecological security. This study built an ecological network using morphological spatial pattern analysis (MSPA), spatial principal component analysis (SPCA) method and minimum cumulative resistance model (MCR) models to analyze the spatial and temporal characteristics and ecological security pattern. Finally, it was optimized by analyzing ecological network indices and using two methods of adding additional ecological sources and stepping stones. The results show that ecological sources and ecological corridors for three phases are located in the central and northern parts with an uneven distribution. In fact, adding new ecological sources is more efficient in balancing the ecological pattern of a study area. The ecological network indices α, β, γ and C values increased by 15.3%, 8.4%, 8.5% and 3.3%, respectively. Constructing and optimizing an ecological network is expected to provide scientific basis for small-scale landscape design, provide theoretical reference for spatial pattern optimization of comprehensive land consolidation projects and coordination of regional development and ecological protection.

Keywords: ecological network; comprehensive land consolidation; small scale; MCR; MSPA; SPCA

Citation: Su, M.; Fang, X.; Sun, K.; Bao, J.; Cao, Y. Construction and Optimization of an Ecological Network in the Comprehensive Land Consolidation Project of a Small Rural Town in Southeast China. *Sustainability* **2023**, *15*, 5984. https://doi.org/10.3390/su15075984

Academic Editor: Sharif Ahmed Mukul

Received: 7 February 2023
Revised: 26 March 2023
Accepted: 28 March 2023
Published: 30 March 2023

1. Introduction

Since China's reform and opening up, the rapid urbanization process has led to a continuous expansion of the urban fringe, with a large amount of agricultural and ecological land being continuously squeezed out, natural habitats disappearing [1]. The Chinese government has carried out relevant land consolidation activities for a while already, from the early period of only focusing on the increase in cultivated land area, to the quantity and quality of farmland, and then to propose to protect the trinity of quantity, quality, and the ecosystem of farmland [2,3]. China's land consolidation has played an important role. In recent decades, with the entry of China into a new phase of high-quality development, the Chinese government has put forward major national development strategies, such as the ecological civilization construction and rural revitalization strategy, and in line with the original land consolidation, has proposed comprehensive land consolidation [4,5]. Comprehensive land consolidation projects require the integration of all elements, including land, capital, livelihood, ecology and industrial benefits [6], to optimize the spatial pattern of production, living, and ecology in rural areas, to increase the area of farmland and the intensive and economical use of land, and to improve the rural habitat. The main contents of comprehensive land consolidation include agricultural land consolidation, rural built-up land consolidation and rural ecological restoration. There are many studies on agricultural land consolidation and built-up land consolidation [7,8]. There are fewer studies on ecological restoration and conservation in the countryside, and in particular, optimization of ecological security patterns in rural areas. Building an ecological network to improve the connectivity of a comprehensive land consolidation area is of great significance to enhance rural ecological security and balance the ecological pattern of the countryside. However, comprehensive land consolidation projects require each project area to add 5% new farmland, and the new farmland index can be used for construction land index transfer, making the local government gain a large amount of revenue [9,10]. Some local governments often adopt the practice of reclaiming some scattered settlements for construction land or reclaiming parkland, forest land and grassland as farmland in order to pursue the maximum benefit of capital [11]. The conflicting objectives of economic income and ecological network protection often lead to the formation of a large scale of farmland and the reduction of biodiversity in practice. Therefore, it is important to explore how to design the landscape ecological pattern of a comprehensive land consolidation project so that its ecological environment is protected with minimal loss of economic benefits.

By increasing the landscape connectivity to re-establish the ecological connectivity of landscape components and strengthen the construction of networked landscape structures, it becomes inevitable to optimize the overall service function of the landscape and maintain regional ecological security. Ecological networks, according to Charles (1990), are "open spaces that connect parks, nature reserves, cultural landscapes, or historic sites and their communities" [12]. From the standpoint of biological conservation, Bennett et al. (2006) argue that ecological networks are continuous regional aggregations of adjacent natural landscape elements that have a significant impact on the survival and reproduction of organisms and must be protected by humans [13]. An ecological network is defined in landscape ecology as an open space that connects ecological patches in the landscape via ecological corridors to form an organic and complete network. The system is used to conserve ecological diversity and maintain the landscape's integrity in order to ensure the landscape's multiple ecological, economic, social, cultural, and aesthetic functions [14,15]. Although scholars' definitions of ecological networks differ, they all emphasize their integrity, connectivity, and ecological service functions.

By connecting high-quality habitats in structure and function, regional-scale ecological networks maintain the stability of the ecological security pattern [16–18]. A stable ecological network not only promotes biodiversity but also provides coordinated economic and ecological development [19,20]. One of the current research hotspots in the context of landscape fragmentation and habitat loss is the construction of ecological networks through landscape connectivity to ensure the ecological security of areas [21–25]. The ecological network is composed of ecological source and ecological corridor. The ecological source is the high-quality human settlement environment, and the ecological corridor is the path connecting the ecological source. The ecological corridor serves many purposes, including transportation and communication, pollution filtering, wind and sand control, and flood regulation. They have the ability to connect fragmented ecological patches and make the urban ecological network system highly connected, both in the landscape and within the city [26]. The development and optimization of ecological networks can significantly improve the service function of regional natural ecosystems and is an important step toward achieving regional sustainable development. It will not only improve biodiversity, aesthetics, and cultural features, but it will also play an important role in the development of sustainable cities [27].

In recent years, landscape connectivity studies have also received widespread attention from scholars [28,29], and landscape index methods [30], morphological spatial pattern analysis (MSPA) [31,32], circuit theory [33], graph theory methods, and minimum cumulative resistance models (MCR) [34] are common methods for constructing connectivity models. Least cost path (LCP) is a commonly used method in connectivity analysis [35]. This method assumes that species' dispersal abilities are dependent on landscape matrix characteristics that facilitate or hinder movement between patches [36]. The shortest distance between habitat patches with the least amount of obstruction is determined using least cost paths. However, because the results represent only a line between two points and not a realistic corridor, they are of limited use for conservation efforts [37]. Researchers have increasingly concentrated on least cost corridors, which represent a cumulative cost gradient, making them more similar to functional areas that connect habitats and thus more realistic in terms of conservation objectives [38].

Graph theory is also an approach in connectivity analysis. Graph theory uses a topological approach to identify patches, corridors, and matrices in a landscape mosaic into nodes, connections and ecological flow relationships between them, reflecting the complex network structure of ecosystems in a simple and intuitive graphical way. The landscape is translated into a graph theory diagram consisting of habitat patches (nodes) that are more or less connected by a network, with the links representing individual dispersal or flow through the landscape [39]. Graph theory is useful because it aids in determining landscape connectivity [40] and the contribution of each individual patch [41]. There is a direct relationship between graph theory and the index of integration of connectivity (*IIC*) and the probability of connectivity (*PC*). The graph theory-based indices clarify the relative connectivity between the optimal habitat area for the focal species and all patches that comprise the entire habitat area, and can be used to prioritize protected areas in order to identify important patch networks for general landscape connectivity, to analyze the impact of individual patch loss and the selection of individual patches that may be connected to corridors, and to assess newly established landscape structures. These metrics have been used to identify the effects of dispersal on focal species and other ecological flows, as well as to assess landscape connectivity in urban green spaces [42,43]. The MSPA identifies landscape patterns by using image-based morphological classification, which has been used in landscape ecology to identify internal and external fragments [44] as well as connecting features of such corridors [45]. The habitat effectiveness index, for instance, completes MSPA [44]. This approach is useful for preserving biodiversity and managing areas, as it contributes to the development of standards for the assessment of ecological design, landscape planning, and biological conservation [46]. In territorial space planning and land management, landscape connectivity is important for conserving biodiversity, maintaining ecosystem stability and integrity, and building ecological security patterns in the landscape, and there is a growing consensus that restoring landscape connectivity not only mitigates landscape fragmentation caused by urbanization and protects the ecological integrity of the region, but is also essential for curbing global climate change [47].

Most of the existing research on landscape connectivity is concerned with the construction and optimization of regional ecological networks [48,49]. However, most existing studies use municipal boundaries or complete geographic units as the study area, using data at large and medium scales. At such scales, the presence of small ecological sources or stepping stones can easily be overlooked, and landscape connectivity cannot be calculated accurately. By selecting a small-scale study area, the impact of stepping stones on the functional connectivity of the study area can be studied in a more detailed way. Furthermore, for small-scale rural landscapes, and as semi-natural ecosystems, it is necessary to take into account not only the disturbing factors of human activity, but also the natural background of the area, such as slope and altitude, which affect geological hazards.

Comprehensive land consolidation is widely promoted in Zhejiang Province, which is advanced and representative compared to other regions in China, and has formed a more mature theoretical system and practical experience. As a traditional agricultural town, the problem of resource depletion and lack of vitality for development is representative of Xiepu Town located in Ningbo City, Zhejiang Province. Xiepu Town fully reflects the development bottlenecks and ecological problems faced by traditional agricultural towns in China in the context of rapid urbanization. At the same time, the study of a typical area can reflect common problems in the existing model of the comprehensive land consolidation project, providing a realistic basis for landscape planning and optimization strategies for comprehensive land consolidation in China. In general, this study focuses on how to optimize the ecological security pattern of comprehensive land consolidation projects under the win–win situation of economy and environment, and provides theories and methods for rural ecological restoration and village-scale landscape planning. The main research objectives are as follows: (1) build comprehensive resistance surface through SPCA and construct the ecological network based on MSPA and MCR model to evaluate ecological security patterns in rural areas; (2) comprehensively analyzed the temporal and spatial characteristics of the ecological network of the study area and existing problems in the comprehensive land consolidation project; (3) build an optimization mechanism of ecological networks for the comprehensive land consolidation project to provide a scientific basis for small-scale landscape design.

2. Materials and Methods

2.1. Study Area

The study area is located in Xiepu Town, Zhenhai District, Ningbo City, Zhejiang Province, at the eastern end of the Ning Shao Plain (121°33′ E to 121°40′ E, 29°59′ N to 30°04′ N). The research area covers Yanshan village, Juedu village and part of Yu Yan village, with the village boundary and Cihai North Road as the boundary, with a total area of 767.30 hm^2. The study area is shown in Figure 1. The study area has a subtropical monsoon climate with long hours of light and abundant rainfall, which provides excellent conditions for agricultural production. Due to its coastal location, it is susceptible to Pacific typhoons in summer. The average annual temperature is 16.3 °C, the average annual sunshine duration is 1944 h, the sunshine rate is 44%, and the average annual precipitation is 1350 mm. The overall topography of the study area is high in the north and low in the south, with the maximum elevation of the low ridge at Yu Yan village on the north side being 142 m. The central part is a large water network plain area, accounting for more than 80% of the total area of the study area. Its ground elevation is about 2 m, and the terrain is flat. Ecological hills, such as Xiaonan Hill and Xiaoling Mountain, are situated around the water system in the central part, at an elevation of about 2–83 m.

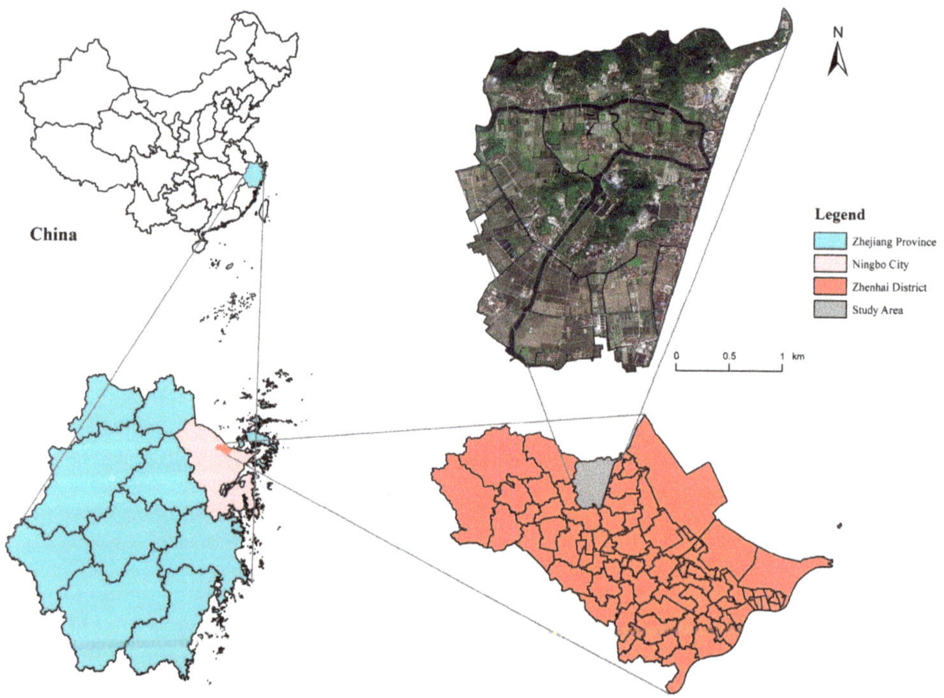

Figure 1. Location of the study area.

The study area includes 17 natural villages, including Tunshan, Jinzhang and Guishan, with a total of 1542 households and a total population of 3736 by the end of 2020. Industries in Xiepu Town are dominated by traditional agriculture and manufacturing, with the tertiary sector being less developed. The total industrial output value of the town above designated size in 2021 totaled CNY 7.733 billion, an increase of 37.8%, respectively, compared to the same period last year. The town's industrial and agricultural output value in 2021 totaled CNY 8 billion, and the per capita disposable income of farmers rose to CNY 36,630, an increase of 18% and 49%, respectively, compared to 2016.

The comprehensive land consolidation project in the study area is carried out in three aspects: farmland consolidation, village consolidation and ecological restoration. Firstly, farmland, as the key land use type in the research area, is the core resource for agricultural development, accounting for 39.78% of the total area of the study area. However, most of the farmland is in between woodland, gardens and industrial land, and the level of aggregation of farmland is low, with farmers operating on a scattered basis, and there is not yet a mature model of transfer operation, which restricts large-scale agricultural production. Therefore, to meet the needs of developing special agriculture and solving existing problems, the comprehensive land consolidation focuses on the goals of food security, economic security, ecological security and social stability, starting from the spatial pattern of farmland, to improve the service function of farmland ecosystem. Secondly, the land use efficiency of the study area is low and the fragmentation of construction land is serious, with a total area of 55 hm^2 of industrial land and a large amount of inefficient construction land. Large-scale factories are basically located around rural settlements, so the demolition and relocation of rural residences and the redistribution of industrial land is undoubtedly the key to village consolidation. Finally, the pressure of ecological environmental protection restoration in Xiepu Town mainly comes from rivers and abandoned mines. The river ecosystem in the town is fragile, with a low self-purification capacity, and is affected by industrial and agricultural production and the daily lives of residents. Under the current spatial layout, the ecological restoration of the river is not enough to solve the river pollution problem. On the other hand, the rapid economic development of Xiepu Town, which relied on mineral resources in its early years, has also caused irreparable damage to the ecological environment. Mining activities have gradually reduced the pH and fertility of the soil, decreasing the soil's carrying capacity and quality, resulting in soil erosion, degradation of natural habitats, reduction of biodiversity and destruction of the ecosystem. In the context of ecological civilization, the transition to the development of a green economy needs to be based on improving agricultural infrastructure, re-planning the layout of industrial land and rural residential bases, and carrying out comprehensive improvement of the living environment.

2.2. Data

The data include: (1) Land use and land cover of the study area in 2013, 2017 and 2021 are obtained from the department of natural resources management. (2) The 30 m × 30 m resolution digital elevation model is derived from the geospatial data cloud (http://www.gscloud.cn, accessed on 11 October 2022). (3) Data on industrial output value, total industrial and agricultural output value, and disposable income of farmers above the scale are obtained from the statistical yearbook of Zhenhai District. (4) The planning objectives and planning paths related to the comprehensive land consolidation project and the current situation of the research area are obtained by the village committee and the engineering design team. (5) The road data are downloaded from the Open Street map.

2.3. Methods

According to the actual research needs of the study area, the land cover in 2013, 2017 and 2021 was divided into seven landscape types: farmland, woodland, garden, grass, water, built-up land, and unused land. At present, the "identify ecological sources—construct resistance surfaces—extract ecological corridors" model is the most widely used method to evaluate landscape connectivity and construct ecological networks. In this study, the ecological network was constructed in four steps and the existing ecological network was optimized (Figure 2). The first step is to identify ecological sources through MSPA and landscape connectivity analysis; the second step is to establish resistance surfaces using the SPCA method; the third step is to extract potential ecological networks using the MCR model and evaluate them using the network analysis method; fourthly, the ecological network was optimized by building stepping stones.

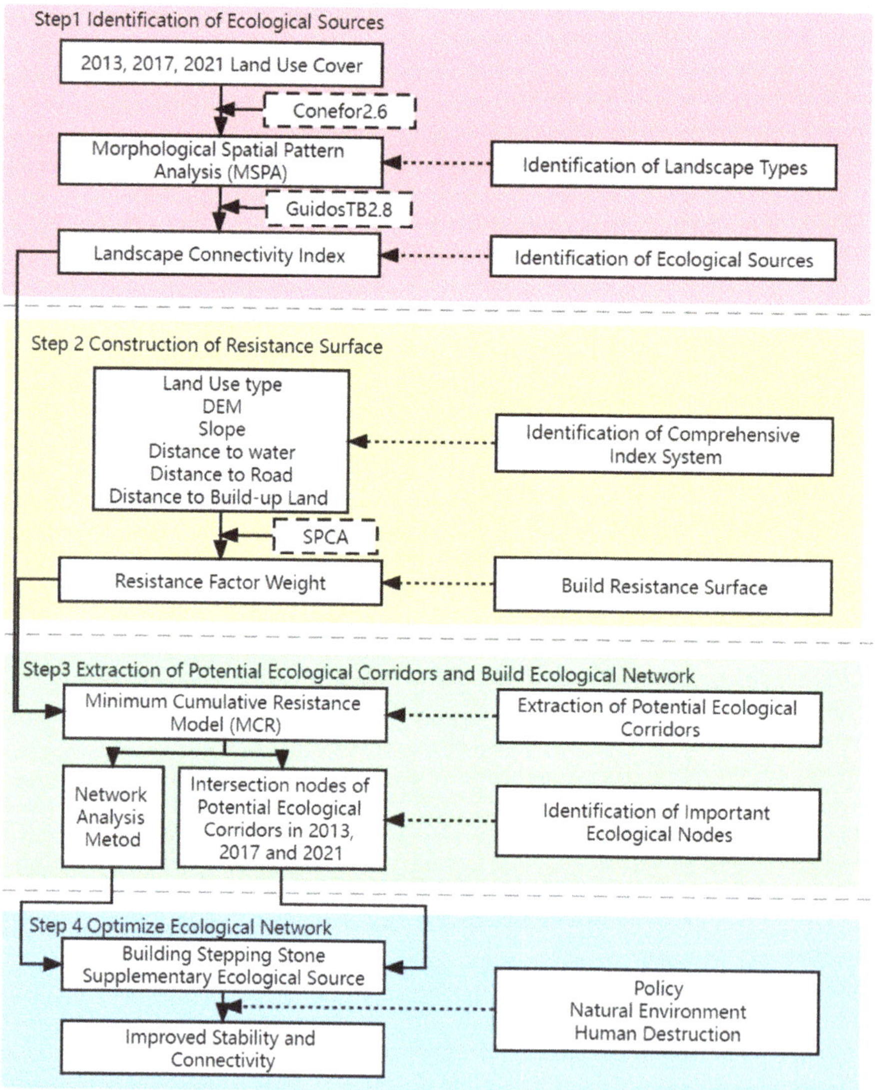

Figure 2. Framework for ecological network construction in the study area.

2.3.1. Identification of Ecological Sources

The theoretical basis of ecological sources is the 'source-sink' theory, which originally referred to existing natural habitats that could be used as a source for the dispersal and maintenance of species [50]. According to the principle of habitat diversity in landscape ecology, patch size is positively related to habitat diversity [51], and therefore ecological source sites should usually be of a certain size.

First of all, based on MSPA to identify ecological source sites [31], landscape elements that have a significant impact on enhancing regional landscape connectivity can be identified. Based on land cover data, seven types of landscape types can be generated, namely core, bridge, edge, loop, perforation, branch and islet. The core areas are the larger habitat patches that provide habitat and migration spaces for species and act as ecological source sites in the ecological network. Soille and Vogt (2009) [52] developed Guidos software for spatial pattern analysis and identification of ecological source sites (http://forest.jrc.ec.europa.eu/download/software/guidos/, accessed on 18 October 2022). To ensure the accuracy of the data, we selected the original land use cover data in order to retain important minor landscape elements. We applied the 8-neighborhood analysis method to the raster data using Guidos analysis software for MSPA analysis. First, woodland and grass were used as foreground (FG) and cultivated land, built-up land, garden land, water, and unused land as background (BG). From the landscape characteristics of the study area, the larger the scale, the more landscape details are missing, while the smaller scale patches are severely fragmented. Moreover, the area of this study area is small, so we identified core area patches larger than 0.5 hm^2 as possible ecological source sites.

In addition, landscape connectivity refers to the process by which the landscape facilitates or hinders the dispersal of species between ecological patches [53,54]. To be precise, the connectivity between ecological patches can be effectively determined from a macroscopic quantitative perspective [55]. After identifying possible ecological sources based on the MSPA method, the landscape connectivity index was used to identify ecological source sites. The core areas were analyzed using the ArcGIS 10.2 platform and Conefor2.6 software. The patch connectivity threshold chosen was 1000 m, and the probability was set to 0.5. The integral index of connectivity (*IIC*), possibility index (*PC*), and patch importance value (*dPC*) are expressed by Equations (1)–(3):

$$IIC = \frac{\sum_{i=1}^{n} \sum_{j=1}^{n} \frac{a_i a_j}{1+nl_{ij}}}{A_L^2}, \quad (1)$$

$$PC = \frac{\sum_{i=1}^{n} \sum_{j=1}^{n} a_i a_j p_{ij}^*}{A_L^2}, \quad (2)$$

$$dPC = \frac{PC - PC_{remove}}{PC} \times 100\%, \quad (3)$$

where, n is the number of patches within the study area; a_i and a_j are the areas of patchs i and j; p_{ij}^* is the maximum probability of species dispersion in patchs i and j; nl_{ij} is the number of connections between patch i and patch j; A_L is the total area of the research range; *IIC* is the overall connectivity index; *PC* is the possible connectivity index of a patch in the landscape, $0 \leq PC \leq 1$, the larger the value of *PC*, the higher the connectivity between the patch and other patches; *dPC* shows the importance of the patch.

2.3.2. Construction of Resistance Surface

A resistance surface is a spatial surface of resistance in a landscape. Landscape resistance was first proposed by Forman (1995) and can be summarized as the impediment to the dispersal of species, material, or energy flows in space [56]. This resistance can be caused by changes in the natural environment or by human activities. The values of resistance vary considerably between landscape types. The success of a species in getting from a source to a target source depends on crossing the resistance distance between the sources [57], meaning the cumulative resistance distance formed by the accumulation of resistance values for each point in the path through which the species crosses a given landscape. Calculating this distance requires a comprehensive resistance surface, so constructing a resistance surface is the basis for extracting ecological network components and is key to ensuring that the ecological network achieves functional connectivity and ecological conservation in the landscape. This study adopts a method based on the ecological safety index, which is constructed from the background characteristics of the ecological environment and the potential threats to it. Commonly used ecological security constraints include elevation, slope, landscape type, vegetation cover, distance from water bodies, distance from roads and distance from residential areas in combination with each ecological safety constraint factor [48,49,58]. The SPCA and expert scoring methods are used to evaluate and classify the ecological safety level of a specific area. A comprehensive resistance surface is also constructed based on the results of the spatial distribution of the ecological safety index evaluation. In this study, a comprehensive resistance index system was constructed based on two resistance factors: natural conditions and human interference (Table 1). The DEM, slope, distance to water bodies and land cover type were selected to represent the resistance factors of natural conditions in the area [49]. The human interference factor is expressed as the distance to the road and built-up land [48].

Table 1. Resistance surface.

Rating Factors	Resistance Factor	Resistance Value				
		1	2	3	4	5
Natural disturbance factors	Land use type	Woodland	Farmland, Grass	Garden land, Unused land	Water	Built-up land
	DEM	<2	2–5	5–15	15–25	>25
	Slope	<15	15–30	30–60	60–90	>90
	Distance to water (m)	0–53	53–130	130–236	236–383	>383
Human disturbance factors	Distance to road (m)	>1306	925–1306	582–925	269–582	0–269
	Distance to built-up land (m)	>343	214–343	121–214	48–121	0–48

In order to avoid factor weight bias caused by subjective influences, this study used SPCA to determine the weights of each factor. With the help of the Principal Components tool of the multivariate analysis function in ArcGIS 10.2 software, the raster data corresponding to each ecological safety evaluation index were input for principal component analysis, and the spatial loadings map corresponding to each principal component and the cumulative contribution rate of each principal component could be obtained. According to the formula for calculating the comprehensive ecological safety index, the variance contribution rate of the principal components obtained from SPCA analysis was calculated to obtain the weights of each evaluation factor. Finally, the evaluation units were weighted and summed, so as to obtain the spatial distribution of ecological safety in the study area. The factors were firstly classified into five categories using the natural breakpoint method, and then the mean values of each breakpoint for 2013, 2017 and 2021 were calculated to derive the classification of each factor for three years. Finally, a reclassification tool was used to assign values to the factors to eliminate inconsistencies in the degree of dimensionality of each factor.

2.3.3. Extract Potential Ecological Corridor

The ecological corridor has many ecosystem service functions, such as biodiversity maintenance, water conservation, soil conservation, flood control and storage. The ecological corridor is a direct channel through which ecological sources communicate with each other and exchange matter and energy, which is conducive to the flow of species between "sources" and matrix. Ecological corridors are also key ecological components to enhance the overall connectivity of the ecosystem. Extraction and construction of ecological corridors can effectively maintain or restore landscape connectivity [59]. On the minimum cumulative resistance surface, the corridor is the resistance trough between two adjacent "sources" and the low-resistance channel that is easiest to contact [60]. MCR extracts the landscape connectivity of the area measured by the resistance distance and calculates the minimum cumulative resistance of the species spreading from the ecological source to a certain point in the space to realize landscape simulation and corridor extraction. This study extracted the ecological corridors between ecological sources that were consistent with the path characteristics of minimum cost [61]. Then, the raster was vectorized, the repeated paths were eliminated, and the vectorized lines were smoothed to determine the spatial position of the ecological corridor with the lowest cumulative resistance value. Potential ecological corridors are extracted based on the MCR model, with Equation (4) as follows:

$$MCR = f_{min}\sum_{j=n}^{i=m} D_{ij} \bullet R_i, \quad (4)$$

where, D_{ij} is the distance from location i to ecological source j; R_i is the resistance encountered in the process of movement; and MCR is the minimum cumulative resistance.

2.3.4. Analyze the Connectivity of Ecological Networks and Build Ecological Network

The network analysis method is widely used in the internal structure of ecological networks, which can be combined with the graph theory method to evaluate the connectivity and complexity of the ecological corridor. By analyzing a graph-theoretic network composed of simplified nodes and connections, the network's topology structure, node connectivity, connection rate and spatial topological relationships were quantitatively described with network closure (α), line-point rate (β), network connectivity (γ) and cost ratio (C) [62,63]. The calculation formulas are defined as follows:

$$\alpha = \frac{L - V + 1}{2V - 5}, \quad (5)$$

$$\beta = \frac{L}{V}, \quad (6)$$

$$\gamma = \frac{L}{3(V - 2)}, \quad (7)$$

$$C = 1 - \frac{1}{d}, \quad (8)$$

where L is the number of corridors; V is the number of ecological sources (number of nodes); d is the total length of the ecological corridor; α index reflects the degree of network circuit, the larger the value indicates that the path provides more possibilities for material and energy flow; β index reflects the number of corridors corresponding to each node, the higher the value, the better the layout of the network; γ index reflects the connectivity of each node in the network, indicating how connected each node is to the other; C represents the input–output relationship, and the lower the value, the more beneficial the ecological network construction.

2.3.5. Optimize Ecological Networks

An ecological network is a complex system of different landscape, forms and types of ecological landscape. In the ecological network based on the MCR model, the distribution of ecological sources and ecological corridors may be unbalanced. In areas without ecological sources, patches with relatively high connectivity can be selected as supplementary ecological sources to effectively solve this problem [40]. Stepping stones provide habitat for species migration, increase landscape connectivity, and promote biodiversity [48,49,64]. Therefore, this research made use of the ecological corridors of 2013, 2017 and 2021, and innovatively integrated the corridors formed by the MCR model, and superimposed the advantageous areas in them to form important ecological nodes. These nodes served as stepping stones to improve the landscape connectivity of the study area.

3. Results

3.1. Land Use Cover Changes

The maps of land use and land cover change in Xiepu Town in 2013, 2017 and 2021 are shown in Figure 3. Farmland is the most widely distributed and spatially interconnected, concentrated within the central area to the large river along the mountain. Woodland is distributed more steadily in the low ridge in the north and in the central area in Xiaonan Hill and Xiaoling Mountain. The garden land is located at the confluence of two rivers. Grass is less distributed in Yuyan village along the side of the mountain road. The area of unused land has decreased significantly, and all is in areas around the mountain that are difficult to develop or have not yet been developed. Built-up land is heavily concentrated in the eastern residential areas and around the various villages.

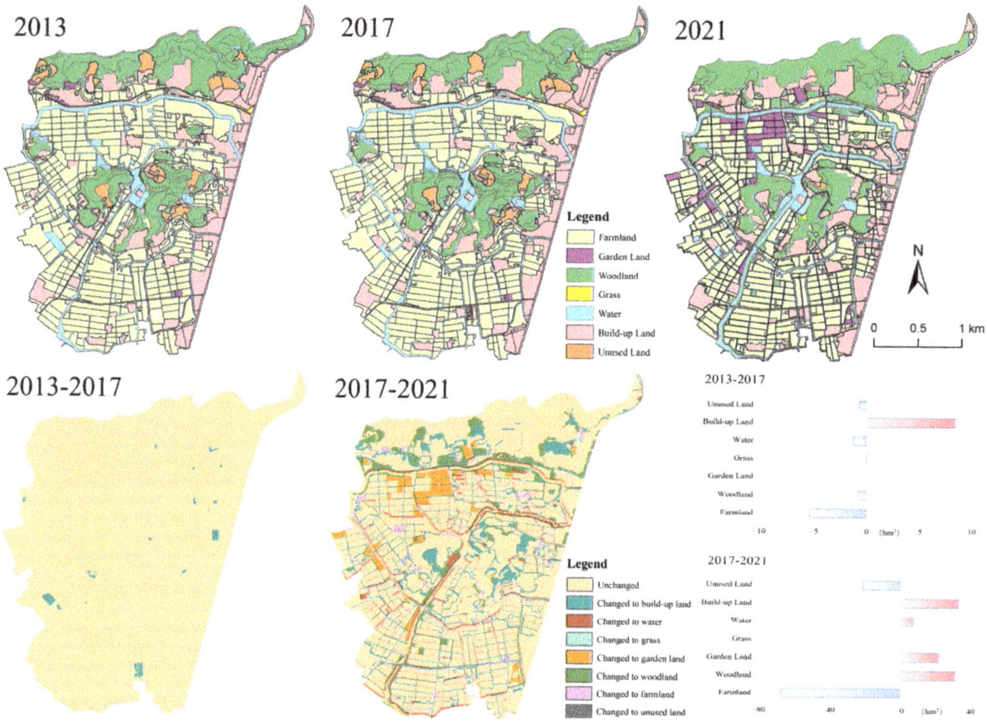

Figure 3. Land use cover and change in the study area from 2013 to 2021.

During the period 2013–2017, spatial changes in land use in Xiepu Town were not obvious, and all that occurred were conversions between other use types and construction land. Spatially, this change was concentrated around North Cihai Road in the east, Luojia village in the south and Fangzhen village in the southwest. It was reflected in marginal expansion, with new construction land coming from the linkage between urban and rural construction land, the occupation of ecological space and the development of unused land. Grassland, forest land and water areas were largely unchanged, and the study area generally had a good basis for agricultural scale as well as rich ecological resources. Garden land remains unchanged, with a fragmented spatial distribution and no scale effect has yet been formed. Cultivated land is of high quality and widely distributed, with large areas of farmland having the potential for large-scale operation.

During the period 2017–2021, the spatial changes in land use in the study were more dramatic, with changes concentrated in the north of the study area and around rivers and roads in the center. In terms of type, construction land was undoubtedly the type that has grown the most, with cultivated land and unused land being the main sources of growth. Farmland has declined to a greater extent in the north, with the emergence of rural roads reducing the concentration of farmland. The large scale of garden land in the north of the study area was a special case, converted from farmland, in addition to the scattering of garden land within the study area. The increase in water body was reflected in the increased width and connectivity of the river channels. The sharp reduction in the amount of unused land was evidence of the fact that Xiepu Town has exploited its reserves for development and the increasing demand for land during this period. In general, the town is undergoing a period of rapid urbanization and the demand for construction land is increasing, while the amount of arable land is declining rapidly in comparison, which is not conducive to the development of large-scale agriculture. When development of unused land is completed, the dynamics changes between various land use types will become more obvious. This also places a higher demand on the comprehensive land consolidation to further improve the land use pattern.

3.2. Ecological Sources Change

In this study, land use data based on the MSPA method, image processing and morphological analysis were used to identify ecological sources in the study area, and at the pixel level, habitat patches that play an important role in landscape connectivity were extracted. The results of ecological sources of the study area are shown in Figure 4. The core area of the study area remained almost unchanged from 2013 to 2017, and showed an upward trend from 2017 to 2021, accounting for less than a quarter of the total area. The core area is concentrated in the central and northern part of the study area. The ecological source of the study area was determined by calculating the *dPC* value of the core area larger than 0.5 hm^2. In 2013 and 2017, 10 regions with *dPC* value greater than 1 were selected as ecological sources. In 2021, 20 ecological sources were selected. These ecological sources are mainly distributed in the central and northern parts of the study area. Among them, the central ecological source has the highest *dPC* value and strong connectivity, which is extremely important in the ecological network. The ecological source of the northern region is mainly composed of woodland area. In 2013, the total core area of the study area was 155.23 hm^2, which decreased by 0.89 hm^2 in 2017. In 2021, the area of core rose to 184.99 hm^2. It can be seen that core patches with good connectivity are mainly distributed in the central and northern parts of the study area, with poor overall connectivity and serious fractures from north to south. Therefore, it is necessary to strengthen the protection and restoration of the southern region, build ecological patches suitable for the survival of species, promote the ecological network of material and energy flow between the northern and southern regions, and promote the healthy and sustainable development of the ecosystem. In addition, after the comprehensive land consolidation, a small amount of woodland has appeared in the southern area of the study area. Therefore, it is necessary to further build the green space and improve the ecological function of the woodland in the study area.

3.3. Resistance Surface and the Change of Factor Weights

Creating a reasonable resistance surface is the basis for establishing an ecological corridor. In this study, six resistance factors were selected and the SPCA method was applied to calculate the weights of each factor for the study area in 2013, 2017 and 2021 (Table 2). Because the land use cover of the study area has changed slightly from 2013 to 2017, the weight of each factor remained the same in both years. Land use type, DEM and distance to road were the three factors with higher weights. In contrast, the land use cover in 2021 shows a significant difference from the previous two years, with a decrease in the amount of cultivated and unused land, and an increase in the amount of other land types to a varying degree. Distance to water and distance to built-up land are also heavily weighted factors.

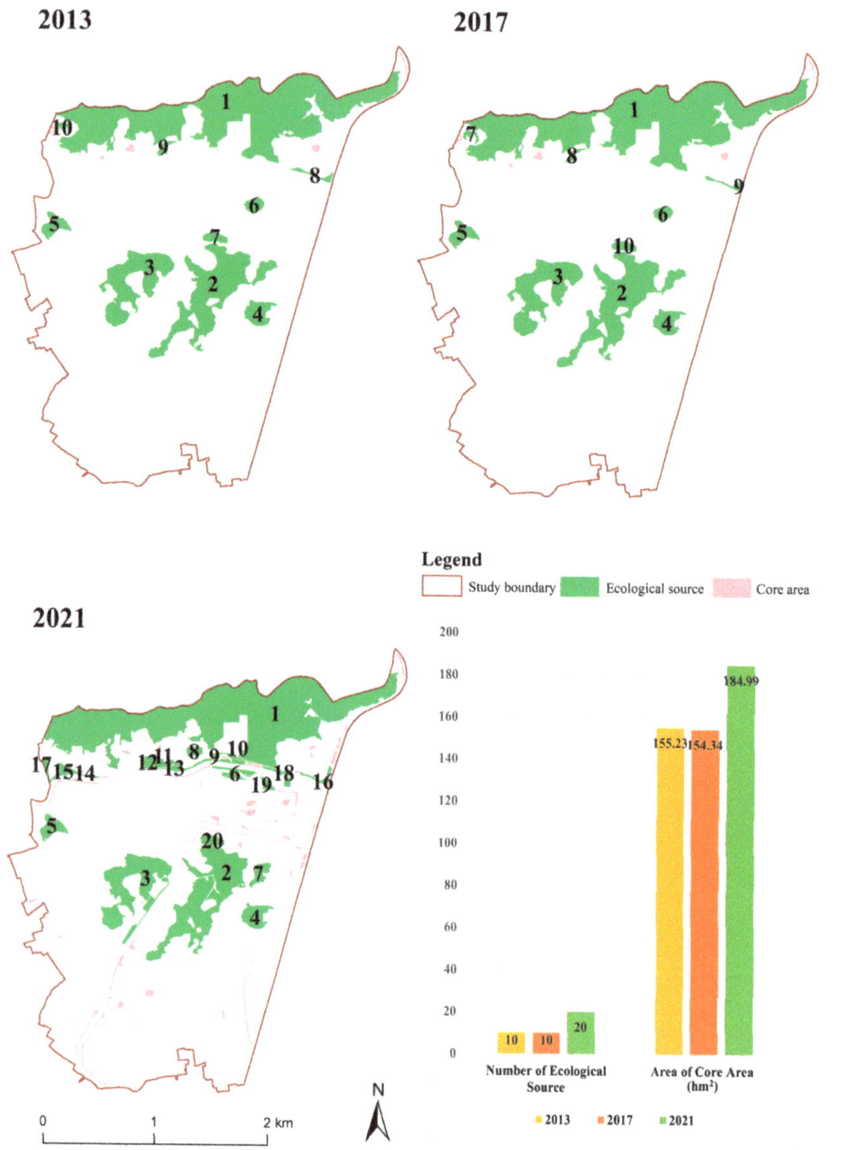

Figure 4. The change of core area and ecological sources area in the study area from 2013 to 2021.

Table 2. Weight of each factor and comprehensive resistance value in 2013, 2017 and 2021.

Index	2013	2017	2021
Land use type	0.47	0.47	0.59
DEM	0.20	0.20	0.02
Slope	0.06	0.06	0.04
Distance to water	0.03	0.03	0.19
Distance to road	0.15	0.15	0.06
Distance to built-up land	0.09	0.09	0.10
Comprehensive resistance	1.3566–4.5869	1.3539–4.5892	0.8078–4.9524

This study chose six resistance factors for the study area in 2013 as an example (Figure 5A–F). As shown in Figure 5A, among the land use type factors, construction land and water have high resistance, which were distributed in all the study area. As shown in Figure 5B–D, pixel with high resistance in DEM factor, slope factor, and distance to water factor are mostly located in the northern and central parts of the research area. Moreover, when the distances from the road and built-up land are further, the resistance in these factors is smaller (Figure 5E,F). The comprehensive resistance surface (Figure 5G) hinders communication between the 10 ecological sources. The lower the resistance value, the greater the transmission and communication in that area. According to the distribution of resistance values, high resistance areas are concentrated in the eastern built-up area and in the north on constructed and unused land. The medium resistance zone is an important area of ecological buffer zone, mainly located in the central mountainous region and the watershed in the study area. The low-resistance areas are mostly arable land that is far from built-up land and roads. This phenomenon mainly reflects the fact that human activities are the most significant factor influencing ecological resistance.

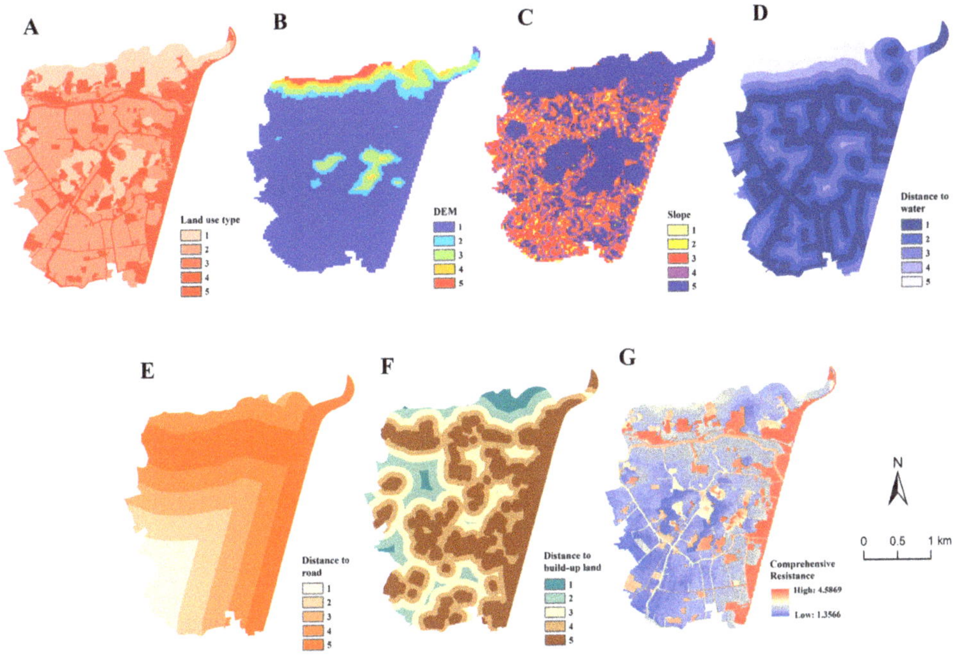

Figure 5. Resistance factor rating in study area in 2013. (**A**): land use type, (**B**): DEM, (**C**): slope, (**D**): distance to water, (**E**): distance to road, (**F**): distance to build-up land, (**G**): comprehensive resistance.

The comprehensive resistance surfaces for the study area in 2013, 2017 and 2021 are shown in Figure 6. The integrated resistance value shows an increasing range of resistance values over time, from 1.36 to 4.59 in 2013, to 0.81 to 4.95 in 2021. The difference in resistance value between pixels in the study area reached its maximum in 2021, indicating that human activities have influenced the ecological resistance of the study area more with the development of eco-tourism and special agriculture in the town of Xiepu. In terms of spatial distribution, all three years of high resistance areas were located in the northern and eastern part of the study area on built-up land, and there were also scattered high resistance areas throughout the whole study area, and the resistance values in these areas increased over time. Resistance values in the center and south-west, which were originally low, have also increased due to the construction of roads and the expansion of built-up land. The rapid urbanization of Xiepu Town has changed the original ecological resistance and the increasing amount of land for construction has challenged the ecological safety pattern of the study area. Therefore, considering the synergistic goal orientation of ecological protection and economic development, there is an urgent need to change the existing land use pattern in the study area, achieve intensive and economical use of construction land, and further promote the ecological effect.

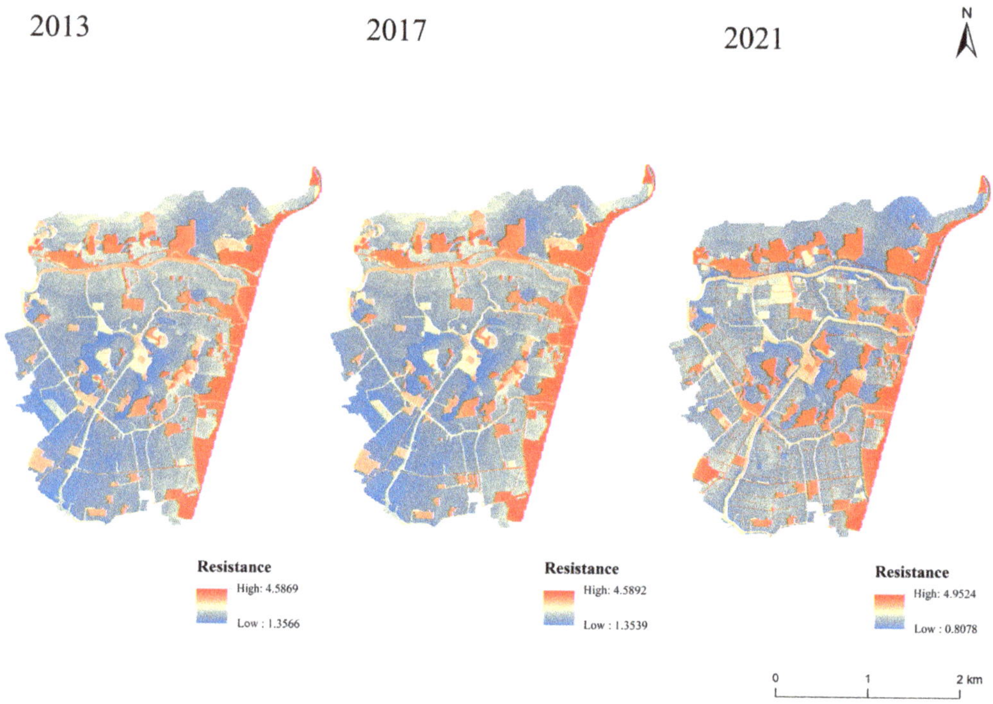

Figure 6. Comprehensive resistance surface in the study area in 2013, 2017 and 2021.

3.4. Establishment and Optimization of Ecological Network

Based on the establishment of a comprehensive resistance surface, the distribution of corridors was identified using the LCP method of the MCR model. Using the ecological source as the source point for the outward spread of the ecological hub, the cost-weighted distance (CWD) and LCP were obtained. After removing some of the redundant corridors, the ecological networks for 2013, 2017 and 2021 were obtained, as shown in Figure 7A–C. As shown in Table 3, there are 10 ecological nodes in 2013 and 2017, while there are 20 ecological nodes in the study area by 2021, showing an increasing trend. At the same time, the number of ecological corridors also rises with the number of ecological sources. β and C indices show different degrees of increase in both time periods. β grows the fastest, by 0.25 from 2013 to 2021, and C index grows by 0.01. α and γ both reach their highest values of 0.93 and 0.96, respectively, in 2017. β is larger indicating that the ecological network of the study area is getting better and better. The ecological network A–C shows a lack of corridor connectivity in the southern and eastern parts of the study area. The eastern part is an agglomeration of villages with high development intensity, a large area of built-up land and limited ecological resources. The southern part of the study area is dominated by agricultural land and has poor ecological functions. Therefore, based on the location and connectivity, core patches should be selected as supplementary ecological sources in the southern part of the study area to balance the ecological source layout.

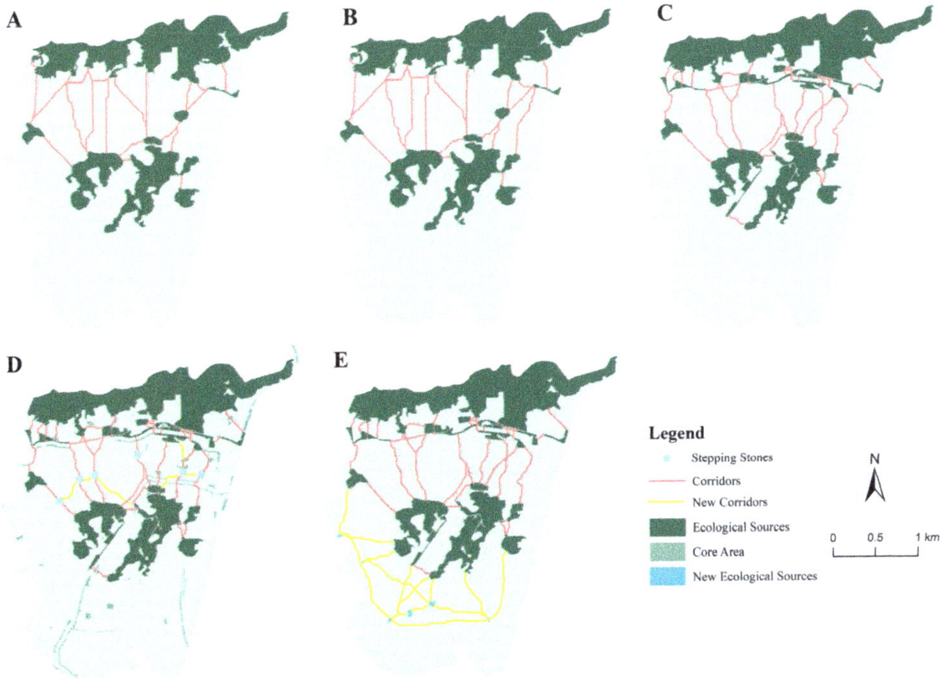

Figure 7. Existing ecological networks and optimized ecological networks ((**A**): ecological network in 2013, (**B**): ecological network in 2017, (**C**): ecological network in 2021, (**D**): stepping stones design based on ecological network in 2013, 2017 and 2021, (**E**): graph showing target ecological sources and corridors).

Table 3. Indices of five ecological networks.

Index	A	B	C	D	E
Corridor number	22	23	49	59	64
Node number	10	10	20	26	26
α	0.87	0.93	0.86	0.72	0.83
β	2.20	2.30	2.45	2.27	2.46
γ	0.92	0.96	0.91	0.82	0.89
C	0.90	0.91	0.91	0.92	0.95
Corridor length (km)	9.695	10.647	11.708	13.158	20.762

In this study, the potential ecological corridors of 2013, 2017 and 2021 were superimposed to identify six intersection points of the potential corridors in each year and set as stepping stones (Figure 7D). The α, β and γ indices of the ecological network with the addition of stepping stones all decreased to different degrees compared to those in 2021. Moreover, as the stepping stones were all located between the central and northern ecological sources in the study area, even if stepping stones were added, they could not constitute an ecological network covering the whole area and could not enhance the landscape connectivity in the southern part of the study area. Therefore, the addition of new ecological source sites may be a better approach. Furthermore, trees have already been planted in the south-western part of the study area, creating a small area of woodland. If the area of the existing small woodland is expanded and upgraded to an ecological source site, the local landscape connectivity can be enhanced more efficiently and constitute a more complete ecological network with a smaller economic loss. This research selected six core patches in the southern and western parts of the study area as supplementary ecological sources and established an ecological network (Figure 7E). α, β, γ and C indices all showed huge increases compared to ecological network D, especially the β and C indices, indicating that the ecological network is becoming more complete and the layout covers a wider area. Spatially, the new ecological nodes have effectively filled the gaps in the west and south, and 9.054 km of new ecological corridors have been added to connect with the original network, forming a new ecological network.

4. Discussion

4.1. Optimization Mechanism of Ecological Network

This study used new framework to establish an optimization mechanism of an ecological network. The network analysis method can be studied through a graphical theory approach, analyzing the quality of ecological patches and corridors, and studying the integrity and stability of ecological networks [65]. In many previous studies, only textual assessments were made based on the optimized networks [49,66]. Although spatially the approach allows for descriptive evaluation of the network, the ecological network indices can be used as quantitative indicators in combination with qualitative studies to enhance the measurement of optimized ecological network connectivity [48,64]. In this study, we first used the ecological network indices to evaluate the connectivity of existing ecological networks. In addition, many studies have pointed out that the addition of stepping stones can have a huge effect on enhancing landscape connectivity in the study area [28,67]. In the process of adding stepping stones in this paper, the ecological networks of 2013, 2017 and 2021 were overlaid, and the intersections of potential corridors in each year were set as stepping stones. Intersections of potential ecological corridors experienced less resistance to getting to the same ecological source site than other pixels, had a greater potential for source-to-source connectivity and dispersal, required fewer barriers and blockages to be crossed, were less costly, and had a greater potential for species to migrate to each other in order to enhance their effect on the flow and transfer of organisms and energy [28]. However, as the ecological source sites in the study area are concentrated in the center and north, this is due to the destruction and blockage of ecological sources by the construction of a large number of towns and agricultural production. As a result, ecological network construction is mainly concentrated in the north. The role of "stepping stones" is not obvious for this study area, and it is the second method to supplement the core area as an ecological source. It is later found that the method of adding stepping stones could not balance the ecological network pattern. Assuming that a small number of existing core areas in the western and southern parts of the study area become new ecological source areas, there is a huge increase in the ecological network indices, so it shows that expanding the existing core areas to optimize them into new ecological source areas is another way to optimize the ecological network. Overall, this study provides an optimization mechanism of "calculating ecological network indices—adding stepping stone—supplying new ecological source" for ecological network construction research.

4.2. The Advantages of Building Ecological Networks for Comprehensive Land Consolidation

Due to the strong scale-dependent regional character of the ecological landscape, especially in areas with high human activity, small-scale areas exhibit significant landscape heterogeneity [68]. Therefore, changes at different scales should be considered in ecological network construction and rural spatial planning [69], and attention should be paid to land-use characteristics and landscape pattern characteristics at smaller scales. This research analyzes the distributional characteristics and connectivity of ecological sources and corridors, down to the small scale. This approach fills a key gap that is lacking in studies at this scale. By identifying ecological sources and extracting potential ecological corridors to form a small-scale ecological network distribution, targeted strategies can be developed for different characteristics of the region and the connectivity of ecological networks in different areas can be improved. Therefore, this study applied two methods of optimizing ecological networks to the study area, adding stepping stones and supplementing new ecological sources. The results show that the addition of new ecological source sites is more effective in improving landscape connectivity and balancing ecological patterns. The "establish ecological network—calculating ecological network indices—adding stepping stones—complementing new ecological sources" approach helps to fulfil the new requirements for the efficient use of natural resources and ecological protection and restoration in the comprehensive land consolidation project with local conditions, and provides a method for improving the ecological pattern and a new approach to territorial space planning and ecological restoration.

4.3. Reconciling Conflicts between Rural Development and Ecological Protection

Ecological sources that require ecological preservation have been identified through the development of the ecological network. As a typical agricultural town, Xiepu Town has little available land for expansion, which limits the amount of space for development in the research area and exacerbates ecological issues. Therefore, a more adaptable land use policy needs to be created to satisfy the needs of both ecological preservation and economic growth. As a comprehensive land consolidation project, the area's goals include maximizing the use of the land's spatial configuration, achieving the harmony of economic, social, and ecological benefits, and fostering the integrated development of urban and rural areas [70]. In order to accomplish these objectives, rural spatial planning and landscape pattern optimization must take into account both the countryside's production function and its ecological function, coordinating both rural growth and ecological protection. As a result, additional new ecological sources from woodlands were chosen for this investigation. Because woodland is a fundamental region in MSPA identification and also has specific ecological services, it may supply more ecological functions, improve landscape connectivity and biodiversity, and balance the ecological pattern when enlarged to become an ecological source. The area that needs to be changed must be less than what would be required to rebuild the ecological source area, and this method can also lessen the economic loss to the study area while increasing the effectiveness of land use and making it easier to coordinate production with the environment's ecological needs.

4.4. Research Limitations and Future Research Directions

The integration of ecological networks and spatial planning is more suited to the protection of regional ecological environments and the coordination of development and conservation than traditional land consolidation approaches. The northern and central regions of the research area make up the majority of the ecological network that was built. In this study, the design of the study area was optimized by the employment of two strategies to modify the geographical distribution of the ecological network. Although the optimization method theoretically adds new ecological nodes, the theoretical solution should be validated by field data in order to make it practicable as well as to monitor, evaluate, and manage the research area adaptively. Whether, for instance, the recently installed stepping stones and ecological source disrupt the present land use and bother the locals with regard to commuting. Additionally, this study does not take into account any particular species and instead focuses on rural ecological land. Land use cover affects ecological processes, species migration, and reproduction, thus future research should take into account the behaviors of specific creatures in the studied area and use additional data from multiple sources for a thorough analysis.

5. Conclusions

This paper provides the spatial and temporal changes in ecological networks in a comprehensive land consolidation project of an agricultural town along the southeast coast of China from 2013 to 2021, and develops a framework for the spatial optimization of ecological networks. Our analysis highlights that the increase in built-up land area and the fragmentation of core areas have a significant influence on ecological security patterns. Meanwhile, it also underlines that the distribution of existing ecological source areas and the specific extent of potential ecological corridors are detrimental to landscape connectivity and affect the integrity and stability of ecological landscapes and planning and management. A better understanding of the role of conserving existing ecological source areas and optimizing existing ecological networks is significant for small-scale rural spatial optimization. Robust evidence discussed in this study suggests that the mechanisms of optimization of ecological networks are critical in facilitating the development of ecological planning and spatial pattern optimization for comprehensive land consolidation. Future research needs to focus on the validation of field data to identify feasible ways to overcome these barriers and make progress towards sustainable development goals in the countryside, and to provide a theoretical basis and scientific reference for rural spatial planning and comprehensive land consolidation in terms of policy and planning methods.

Author Contributions: Conceptualization, M.S., X.F. and Y.C.; Methodology, M.S. and X.F.; Software, M.S.; Formal analysis, M.S.; Investigation, K.S. and J.B.; Resources, Y.C.; Data curation, M.S., K.S. and J.B.; Writing—original draft, M.S.; Writing—review & editing, X.F. and Y.C.; Visualization, M.S.; Supervision, Y.C. All authors have read and agreed to the published version of the manuscript.

Funding: This research was funded by the Key Project of the National Social Science Fund of China (No. 20AGL025).

Institutional Review Board Statement: Not applicable.

Informed Consent Statement: Not applicable.

Data Availability Statement: The data used to support the findings of this study are available from the corresponding author upon request.

Conflicts of Interest: The authors declare no conflict of interest.

References

1. Yonaba, R.; Biaou, A.C.; Koïta, M.; Tazen, F.; Mounirou, L.A.; Zour'e, C.O.; Queloz, P.; Karambiri, H.; Yacouba, H. A dynamic land use/land cover input helps in picturing the Sahelian paradox: Assessing variability and attribution of changes in surface runoff in a Sahelian watershed. *Sci. Total Environ.* **2021**, *757*, 143792. [CrossRef] [PubMed]
2. Xia, F.Z.; Yang, Y.M.; Yan, J.M. Review on the connotation of China's comprehensive land consolidation in the past 40 years: Stage evolution and development change. *Chin. Land Sci.* **2018**, *5*, 78–85.
3. Xu, H.Z. Mechanism and implementation path of comprehensive land consolidation to boost rural revitalization. *Guizhou Soc. Sci.* **2021**, *5*, 144–152. [CrossRef]
4. Wu, C.F.; Xiao, W.; Cao, Y.; Fang, K. *Ecological Restoration of Territorial Space*; Geological Publishing House: Beijing, China, 2019; pp. 304–305.
5. Yan, J.M.; Xia, F.Z.; Li, Q. Top-level design of comprehensive land consolidation strategy in China. *Trans. Chin. Soc. Agric. Eng.* **2012**, *28*, 1–9.
6. Li, Y.X.; Li, J.; Chen, H.; Wang, Z.J. Landscape connectivity evaluation and temporal-spatial characteristics of Guiyang City from 2008 to 2017 based on MSPA and MCR models. *Chin. J. Ecol.* **2022**, *41*, 1240–1248.
7. Tang, H.; Yun, W.; Liu, W.; Sang, L. Structural Changes in the Development of China's Farmland Consolidation in 1998–2017: Changing Ideas and Future Framework. *Land Use Policy* **2019**, *89*, 104212. [CrossRef]
8. Zhang, M.; He, T.; Wu, C.; Li, G. The Spatiotemporal Changes in Ecological–Environmental Quality Caused by Farmland Consolidation Using Google Earth Engine: A Case Study from Liaoning Province in China. *Remote Sens.* **2022**, *14*, 3646. [CrossRef]
9. Kong, X.S.; Wang, J.; Jin, Z.F.; Er, L.L. Rural land consolidation transformation and innovative thinking for rural revitalization. *Chin. Land Sci.* **2019**, *5*, 95–102.
10. Xiao, W.; Hou, L.; Yue, W.Z. Connotation, dilemma and countermeasure of comprehensive land consolidation. *Chin. Land* **2022**, *7*, 12–15. [CrossRef]
11. Li, J.L.; Xu, J.G.; Chu, J.L. Research on Urban Ecological Security Pattern Based on Circuit Theory: A Case Study of Anqing City. *Resour. Environ. Yangtze Val.* **2020**, *29*, 1812–1824.
12. Charles, E.L. *Greenway for America*; Johns Hopkins University: Baktimore, MD, USA, 1990; p. 1.
13. Bennett, A.F.; Radford, J.Q.; Haslem, A. Properties of Land Mosaics: Implications for Nature Conservation in Agricultural Environments. *Biol. Conserv.* **2006**, *133*, 250–264. [CrossRef]
14. Xing, L.; Hu, M.; Wang, Y. Integrating Ecosystem Services Value and Uncertainty into Regional Ecological Risk Assessment: A Case Study of Hubei Province, Central China. *Sci. Total Environ.* **2020**, *740*, 140126. [CrossRef] [PubMed]
15. Sun, X.; Crittenden, J.C.; Li, F.; Lu, Z.; Dou, X. Urban Expansion Simulation and the Spatio-Temporal Changes of Ecosystem Services, a Case Study in Atlanta Metropolitan Area, USA. *Sci. Total Environ.* **2018**, *622–623*, 974–987. [CrossRef] [PubMed]

16. Yu, Q.; Yue, D.; Wang, Y.; Kai, S.; Fang, M.; Ma, H.; Zhang, Q.; Huang, Y. Optimization of Ecological Node Layout and Stability Analysis of Ecological Network in Desert Oasis: A Typical Case Study of Ecological Fragile Zone Located at Deng Kou County (Inner Mongolia). *Ecol. Indic.* **2018**, *84*, 304–318. [CrossRef]
17. Li, X.; Lao, C.; Liu, Y.; Liu, X.; Chen, Y.; Li, S.; Ai, B.; He, Z. Early Warning of Illegal Development for Protected Areas by Integrating Cellular Automata with Neural Networks. *J. Environ. Manag.* **2013**, *130*, 106–116. [CrossRef]
18. Baggio, J.A.; Hillis, V. Managing ecological disturbances: Learning and the structure of social-ecological networks. *Environ. Modell. Softw.* **2018**, *109*, 32–40. [CrossRef]
19. Hostetler, M.; Allen, W.; Meurk, C. Conserving Urban Biodiversity? Creating green Infrastructure Is Only the First Step. *Landsc. Urban Plan.* **2011**, *100*, 369–371. [CrossRef]
20. Lookingbill, T.R.; Gardner, R.H.; Ferrari, J.R.; Keller, C.E. Combining a dispersal model with network theory to assess habitat connectivity. *Ecol. Appl.* **2010**, *20*, 427–441. [CrossRef]
21. Chen, C.; Shi, L.; Lu, Y.; Yang, S.; Liu, S. The Optimization of Urban Ecological Network Planning Based on the Minimum Cumulative Resistance Model and Granularity Reverse Method: A Case Study of Haikou, China. *IEEE Access* **2020**, *8*, 43592–43605. [CrossRef]
22. Guo, L.; Luo, J.; Yuan, M.; Huang, Y.; Shen, H.; Li, T. The Influence of Urban Planning Factors on PM2.5 Pollution Exposure and Implications: A Case Study in China Based on Remote Sensing, LBS, and GIS Data. *Sci. Total Environ.* **2019**, *659*, 1585–1596. [CrossRef]
23. Vuilleumier, S.V.; Prélaz-Droux, R. Map of ecological networks for landscape planning. *Landsc. Urban Plan.* **2002**, *58*, 157–170. [CrossRef]
24. Battisti, C. Habitat fragmentation, fauna and ecological network planning: Toward a theoretical conceptual framework. *Ital. J. Zool.* **2003**, *70*, 241–247. [CrossRef]
25. Hüse, B.; Szabo, S.; Deák, B.; Tóthmérész, B. Mapping an ecological network of green habitat patches and their role in maintaining urban biodiversity in and around Debrecen city (Eastern Hungary). *Land Use Policy* **2016**, *57*, 574–581. [CrossRef]
26. Tang, Y.; Gao, C.; Wu, X. Urban Ecological Corridor Network Construction: An Integration of the Least Cost Path Model and the InVEST Model. *Int. J. Geo-Inf.* **2020**, *9*, 33. [CrossRef]
27. Ignatieva, M.; Stewart, G.H.; Meurk, C. Planning and Design of Ecological Networks in Urban Areas. *Landsc. Ecol.* **2011**, *7*, 17–25. [CrossRef]
28. Luo, Y.; Wu, J.; Wang, X.; Peng, J. Using Stepping-Stone Theory to Evaluate the Maintenance of Landscape Connectivity under China's Ecological Control Line Policy. *J. Clean. Prod.* **2021**, *296*, 126356. [CrossRef]
29. Heintzman, L.J.; McIntyre, N.E. Assessment of Playa Wetland Network Connectivity for Amphibians of the South-Central Great Plains (USA) Using Graph-Theoretical, Least-Cost Path, and Landscape Resistance Modelling. *Landsc. Ecol.* **2021**, *36*, 1117–1135. [CrossRef]
30. Yan, X.; Niu, Z.G. Study on wetland connectivity in Baiyangdian Basin. *Acta Ecol. Sin.* **2019**, *39*, 9200–9210.
31. Ye, H.; Yang, Z.; Xu, X. Ecological Corridors Analysis Based on MSPA and MCR Model—A Case Study of the Tomur World Natural Heritage Region. *Sustainability* **2020**, *12*, 959. [CrossRef]
32. Wickham, J.D.; Riitters, K.H.; Wade, T.G.; Vogt, P. A national assessment of green infrastructure and change for the conterminous United States using morphological image processing. *Landsc. Urban Plan.* **2010**, *94*, 186–195. [CrossRef]
33. Li, T.; Gong, Y.B.; Ge, J.Z.; Qi, Z.X.; Xie, S.B. Construction of urban landscape ecological security pattern based on Circuit theory: A case study of Hengyang City, Hunan Province. *Chin. J. Appl. Ecol.* **2021**, *32*, 2555–2564.
34. Li, H.H.; Ma, T.K.; Wang, K.; Tan, M.; Qu, J.F. Construction of Ecological Security Pattern in Northern Peixian Based on MCR and SPCA. *J. Ecol. Rural Environ.* **2020**, *36*, 1036–1045.
35. Correa Ayram, C.A.; Mendoza, M.E.; Etter, A.; Salicrup, D.R.P. Habitat Connectivity in Biodiversity Conservation: A Review of Recent Studies and Applications. *Prog. Phys. Geogr. Earth Environ.* **2016**, *40*, 7–37. [CrossRef]
36. Adriaensen, F.; Chardon, J.P.; De Blust, G.; Swinnen, E.; Villalba, S.; Gulinck, H.; Matthysen, E. The Application of 'Least-Cost' Modelling as a Functional Landscape Model. *Landsc. Urban Plan.* **2003**, *64*, 233–247. [CrossRef]
37. Rudnick, D.; Ryan, S.; Beier, P. The role of landscape connectivity in planning and implementing conservation and restoration priorities. *Issues Ecol.* **2012**, *16*, 20.
38. Beier, P.; Spencer, W.; Baldwin, R.F.; McRAE, B.H. Toward Best Practices for Developing Regional Connectivity Maps: Regional Connectivity Maps. *Conserv. Biol.* **2011**, *25*, 879–892. [CrossRef] [PubMed]
39. Bunn, A.G.; Urban, D.L.; Keitt, T.H. Landscape Connectivity: A Conservation Application of Graph Theory. *J. Environ. Manag.* **2000**, *59*, 265–278. [CrossRef]
40. Laita, A.; Kotiaho, J.S.; Mönkkönen, M. Graph-Theoretic Connectivity Measures: What Do They Tell Us about Connectivity? *Landsc. Ecol.* **2011**, *26*, 951–967. [CrossRef]
41. Rubio, L.; Saura, S. Assessing the importance of individual habitat fragments as irreplaceable connecting elements: An analysis of simulated and real landscape data. *Ecol. Complex.* **2012**, *11*, 28–37. [CrossRef]
42. Decout, S.; Manel, S.; Miaud, C.; Luque, S. Integrative Approach for Landscape-Based Graph Connectivity Analysis: A Case Study with the Common Frog (Rana Temporaria) in Human-Dominated Landscapes. *Landsc. Ecol.* **2012**, *27*, 267–279. [CrossRef]
43. Ziółkowska, E.; Ostapowicz, K.; Kuemmerle, T.; Perzanowski, K.; Radeloff, V.C.; Kozak, J. Potential Habitat Connectivity of European Bison (Bison Bonasus) in the Carpathians. *Biol. Conserv.* **2012**, *146*, 188–196. [CrossRef]

44. Vogt, P.; Riitters, K.H.; Estreguil, C.; Kozak, J.; Wade, T.G.; Wickham, J.D. Mapping Spatial Patterns with Morphological Image Processing. *Landsc. Ecol.* **2007**, *22*, 171–177. [CrossRef]
45. Vogt, P.; Riitters, K.H.; Iwanowski, M.; Estreguil, C.; Kozak, J.; Soille, P. Mapping Landscape Corridors. *Ecol. Indic.* **2007**, *7*, 481–488. [CrossRef]
46. Clerici, N.; Vogt, P. Ranking European Regions as Providers of Structural Riparian Corridors for Conservation and Management Purposes. *Int. J. Appl. Earth Obs. Geoinf.* **2013**, *21*, 477–483. [CrossRef]
47. He, K.; Lin, T.; Wu, J.F.; Sui, M.F.; Liu, L.; Ding, G.C. Construction of green infrastructure network in Fuzhou City Center based on spatial priority. *Chin. J. Appl. Ecol.* **2021**, *32*, 1424–1432.
48. Nie, W.; Shi, Y.; Siaw, M.J.; Yang, F.; Wu, R.; Wu, X.; Zheng, X.; Bao, Z. Constructing and Optimizing Ecological Network at County and Town Scale: The Case of Anji County, China. *Ecol. Indic.* **2021**, *132*, 108294. [CrossRef]
49. Zhang, J.; Zhu, H.; Zhang, P.; Song, Y.; Zhang, Y.; Li, Y.; Rong, T.; Liu, Z.; Yang, D.; Lou, Y. Construction of GI Network Based on MSPA and PLUS Model in the Main Urban Area of Zhengzhou: A Case Study. *Front. Environ. Sci.* **2022**, *10*, 878656. [CrossRef]
50. Yu, K.J. Landscape ecological security pattern of biological conservation. *Acta Ecol. Sin.* **1999**, *19*, 8–15.
51. Guo, X.; Huang, G.; Jia, P.; Wu, J. Estimating Fine-Scale Heat Vulnerability in Beijing Through Two Approaches: Spatial Patterns, Similarities, and Divergence. *Remote Sens.* **2019**, *11*, 2358. [CrossRef]
52. Soille, P.; Vogt, P. Morphological Segmentation of Binary Patterns. *Pattern Recognit. Lett.* **2009**, *30*, 456–459. [CrossRef]
53. Taylor, P.D.; Fahrig, L.; Henein, K.; Merriam, G. Connectivity Is a Vital Element of Landscape Structure. *Oikos* **1993**, *68*, 571. [CrossRef]
54. Forman, R.T.T.; Collinge, S.K. Nature Conserved in Changing Landscapes with and without Spatial Planning. *Landsc. Urban Plan.* **1997**, *37*, 129–135. [CrossRef]
55. Clergeau, P.; Burel, F. The role of spatio-temporal patch connectivity at the landscape level: An example in a bird distribution. *Landsc. Urban Plan.* **1997**, *38*, 37–43. [CrossRef]
56. Forman, R.T.T. Some General Principles of Landscape and Regional Ecology. *Landscape Ecol.* **1995**, *10*, 133–142. [CrossRef]
57. Knaapen, J.P.; Scheffer, M.; Harms, B. Estimating Habitat Isolation in Landscape Planning. *Landsc. Urban Plan.* **1992**, *23*, 1–16. [CrossRef]
58. Keeley, A.T.H.; Beier, P.; Gagnon, J.W. Estimating landscape resistance from habitat suitability: Effects of data source and nonlinearities. *Landsc. Ecol.* **2016**, *31*, 2151–2162. [CrossRef]
59. Peng, J.; Li, H.L.; Liu, Y.X.; Hu, Y.N.; Yang, Y. Ecological security pattern identification and optimization strategy in Xiong'an New Area. *Acta Geo. Sin.* **2018**, *73*, 701–710.
60. Luo, Y.; Wu, J. Linking the Minimum Spanning Tree and Edge Betweenness to Understand Arterial Corridors in an Ecological Network. *Landsc. Ecol.* **2021**, *36*, 1549–1565. [CrossRef]
61. McRae, B.H.; Dickson, B.G.; Keitt, T.H.; Shah, V.B. Using Circuit Theory to Model Connectivity in Ecology, Evolution, and Conservation. *Ecology* **2008**, *89*, 2712–2724. [CrossRef]
62. Zhang, Z.; Meerow, S.; Newell, J.P.; Lindquist, M. Enhancing Landscape Connectivity through Multifunctional Green Infrastructure Corridor Modeling and Design. *Urban For. Urban Green.* **2019**, *38*, 305–317. [CrossRef]
63. Zhao, S.; Ma, Y.; Wang, J.; You, X. Landscape Pattern Analysis and Ecological Network Planning of Tianjin City. *Urban For. Urban Green.* **2019**, *46*, 126479. [CrossRef]
64. Huang, X.; Wang, H.; Shan, L.; Xiao, F. Constructing and Optimizing Urban Ecological Network in the Context of Rapid Urbanization for Improving Landscape Connectivity. *Ecol. Indic.* **2021**, *132*, 108319. [CrossRef]
65. Xiao, H.; Liu, Y.H.; Yu, Z.R.; Zhang, Q.; Zhang, X. Combination of ecoprofile and least-cost model for eco-network planning. *Chin. Geogr. Sci.* **2014**, *24*, 113–123. [CrossRef]
66. Hu, C.G.; Wang, Z.Y.; Wang, Y.; Sun, D.Q.; Zhang, J.X. Combining MSPA-MCR Model to Evaluate the Ecological Network in Wuhan, China. *Land* **2022**, *11*, 213. [CrossRef]
67. Huang, X.F.; Wu, C.F.; You, H.Y.; Xiao, W.; Zhong, S.Q. Construction of rural landscape ecological corridor in water network plain area based on MCR Model. *Trans. Chin. Soc. Agric. Eng.* **2019**, *35*, 243–251.
68. Shi, F.; Liu, S.; Sun, Y.; An, Y.; Zhao, S.; Liu, Y.; Li, M. Ecological Network Construction of the Heterogeneous Agro-Pastoral Areas in the Upper Yellow River Basin. *Agric. Ecosyst. Environ.* **2020**, *302*, 107069. [CrossRef]
69. Gonthier, D.J.; Ennis, K.K.; Farinas, S.; Hsieh, H.-Y.; Iverson, A.L.; Batáry, P.; Rudolphi, J.; Tscharntke, T.; Cardinale, B.J.; Perfecto, I. Biodiversity Conservation in Agriculture Requires a Multi-Scale Approach. *Proc. R. Soc. B* **2014**, *281*, 20141358. [CrossRef]
70. Ge, D.Z.; Long, H.L. Rural spatial governance and urban-rural integration development. *Acta Geogr. Sin.* **2020**, *75*, 1272–1286.

sustainability MDPI

Article

Tibetan Herders' Life Satisfaction and Determinants under the Pastureland Rehabilitation Program: A Case Study of Maduo County, China

Cuizhen Xia [1,2], Lihua Zhou [1,2,*], Ya Wang [3] and Xiaodong Pei [1,2]

1 Institutes of Science and Development, Chinese Academy of Sciences, Beijing 100190, China; xiacuizhen@lzb.ac.cn (C.X.); xdpeiucas@126.com (X.P.)
2 School of Public Policy and Management, University of Chinese Academy of Sciences, Beijing 100049, China
3 Northwest Institute of Eco-Environment and Resources, Chinese Academy of Sciences, Lanzhou 730000, China; wy_2005@yeah.net
* Correspondence: lhzhou@casisd.cn

Abstract: The Pastureland Rehabilitation Program (PRP) has been implemented for nearly 20 years, and the lives of herders in different regions have been affected to varying degrees. The level of people's well-being could measure the success of policy. Taking Maduo County as an example, the life satisfaction of 266 Tibetan herders was investigated through a participatory questionnaire survey in July 2021. Multiple linear regression and optimal scale regression were constructed to analyze the impact of the satisfaction of different aspects of life and PRP on life satisfaction, respectively. The results show that the herders in Maduo County had relative high satisfaction in various aspects of life and overall life, and leisure and consumption were important aspects influencing life satisfaction. The relative living standard, the difficulty in borrowing and the quality of government services had a significant positive impact on herders' life satisfaction. Occupation and migration location also led to the significant differences of life satisfaction. However, the income changes caused by PRP and subsidy levels did not show a significant impact. We discussed the particularity of Maduo County and suggested that more attention should be paid to the improvement of the social environment such as wealth disparity, channels of assistance, working environment, and the effective assistance formulated according to the demands of different groups needs to be optimized continuously, so as to enhance the self-development ability of herders.

Keywords: life satisfaction; herder; Maduo; Pastureland Rehabilitation Program; subjective well-being

Citation: Xia, C.; Zhou, L.; Wang, Y.; Pei, X. Tibetan Herders' Life Satisfaction and Determinants under the Pastureland Rehabilitation Program: A Case Study of Maduo County, China. *Sustainability* **2022**, *14*, 2381. https://doi.org/10.3390/su14042381

Academic Editor: Yu Cao

Received: 31 December 2021
Accepted: 10 February 2022
Published: 19 February 2022

1. Introduction

The pursuit of well-being is as old as human history [1]. Life satisfaction is one of the most important human values [2], an important part of subjective well-being [3,4], and a powerful supplement to the evaluation of an individual's actual state in addition to objective economic and social indicators [5,6]. Its reliability as an approximate estimate of personal utility has been verified in decades of research [7]. Individual well-being today has been also regarded as an indicator of social progress [8,9], and using people's self-reported life satisfaction to assess the effectiveness of governance has been prevalent around the world [10,11]. Life satisfaction surveys have been often used for various groups in different regions, and the reasons for the differences in satisfaction have been also deeply analyzed to find ways to improve human well-being [12–16]. Many studies have shown that in addition to people's own characteristics, social and economic factors can also affect well-being, which may be influenced by policy makers [9].

As one of the typically comprehensive ecological restoration projects in China, the Pastureland Rehabilitation Program (PRP) was started in 2003 and covered almost all grassland areas in China, which account for 41.7% of the country's land area [17]. The project aimed to curb grassland degradation and restore the function of grassland ecosystem through fence construction, resowing, grazing prohibition, rest grazing, rotational grazing, ecological compensation and other measures. However, due to the strict restrictions on the use of grassland, the lives of herders whose livelihood rely heavily on grassland have been seriously changed, which further affects their well-being [18]. The policy support and the adaptability of herders jointly determine the impact of the PRP on the overall well-being of herders' families, and the loss of well-being often leads to many illegal grazing behaviors, such as grazing in grazing-prohibited areas or times, or maintaining a number of livestock exceeding the specified standard [19,20]. Ensuring the well-being of herders is considered to be a solution to alleviate the contradictions between herders and government and promote ecological protection [21,22].

The Three-River Headwaters Region, located in the interior of the Qinghai–Tibetan Plateau, has a particularly critical ecological status because it breeds the Yangtze River, Yellow River, and Lancang River, the three most important rivers to China and Southeast Asian countries. There, alpine grasslands and meadows are the main vegetation types, and nomadism has always been the traditional livelihood of local Tibetans. The central government always gives priority to the ecological quality of the region. Therefore, in the early stage of PRP, a large area of grazing prohibition and ecological immigration were implemented here. How to properly resettle local herders and ensure their livelihood has become a key issue that the government must consider. It has been nearly 20 years since the project was implemented; what is the living condition of the herders in the area, and is their well-being guaranteed? What other changes can the government make to improve well-being? With Maduo County, the source of the Yellow River, as a typical area, we investigated the satisfaction of local herders in various aspects of life and overall life satisfaction through field research, and used the multiple linear regression and optimal scale regression models to analyze the determinants of life satisfaction. Through the identification of the impact path of PRP on the herders' life satisfaction, the optimization suggestions of the program are put forward to improve herders' well-being.

2. Literature Review

Life satisfaction is a subjective evaluation of the life quality made by individuals based on their own standards [23,24]. Previous studies have indicated that recent experiences (accidents, death in family and other emergencies), factors of short and medium term (emotional fluctuations, housing, and marital status, etc.) and very long-term factors (personality and emotional intelligence, etc.) will affect self-reported life satisfaction [6]. Due to the differences of each person's personality, experience, and external environment, the research results on the determinants of life satisfaction are rich and complicated. Economic factors such as income and relative income, employment, and noneconomic or underlying factors such as age, gender, marital, health, education, religion, and social and political institutions have been found to be significantly associated with life satisfaction [25,26].

Three theories could be used to explain the differences in satisfaction, namely need and goal satisfaction theories, process or activity theories, genetic and personality predisposition theories [26,27]. The first theory states that people have high life satisfaction when their needs are met. According to Maslow's Hierarchy of Needs theory, people first need to meet physiological needs, which require adequate availability of basic needs such as food, housing, fuel, etc. [28]. Multiple comparisons theory also offers another explanation of life satisfaction [29]. This theory argues that life satisfaction depends on the individual's comparison with various standards, which may be related to social standards, the level of people around, personal desires and goals, and past experiences [6,30].

In ecologically fragile areas, the guarantee of residents' well being has become more complicated due to the pursuit of ecological protection. In the implementation area of PRP, the external environment and herders' experience have changed under the intervention of various measures, which may further affect herders' perception and life satisfaction (shown in the Table 1). Grazing prohibition has caused herders to lose their available pastures, and forage–livestock balance has stipulated the maximum number of livestock that herders can raise and the time available for grazing. After immigration, herders without other skills have found it difficult to find new jobs, resulting in their increased dependence on government subsidies. In the past 20 years, household income, livelihood adjustments and adaptation, and compensation standard under the PRP have received extensive attention from scholars [31–34].

Table 1. Measures and impact on herders of Pastureland Rehabilitation Program [18,19,35,36].

Measure	Specific Measure	Impact on Herders
Main measures	Grazing prohibition	Pasture would be forbidden to use
	Forage–livestock balance	The number of livestock and grazing time would be limited
	Restoration of degraded grassland	Degraded grassland shall be restored by government
Supporting measures	Fence construction	Each household's pasture would be spatially separated
	Livestock shed construction	Livestock could be raised in sheds
	Artificial grassland construction	The shortage of forage could be alleviated
	Ecological migration	Herders in key protected area would be relocated
	Ecological compensation	The economic losses of herders would be partially compensated

In the implementation area of the PRP, such as Ningxia, Gansu, Inner Mongolia, Qinghai, and other places, many studies have focused on the satisfaction of herders with individual measures such as ecological compensation and grazing prohibition. The results have showed that the policy satisfaction of herders and influencing factors in different regions were different [21,22]. When considering the influencing factors, scholars tend to choose variables from personal characteristics, family capital, environmental factors, policy perception and so on. In terms of objective factors, herders' family size, education level, livestock number, income and expenditure, and social frequency significantly may significantly affect farmers' or herders' policy satisfaction; in terms of subjective factors, relative income, ecological protection awareness, environmental satisfaction, social welfare satisfaction and subjective evaluation of compensation standard may significantly affect farmers' or herders' policy satisfaction [37,38]. Although individual policy satisfaction of herders may further affect their overall well-being [39], focusing on a single measure may not fully reflect the PRP's impact on well-being, and the evaluation of satisfaction with a single measure may also contain the impacts of other measures. Therefore, taking the overall life satisfaction of the herders under the background of PRP as the research object, and determining the impact factors according to the way the grazing-returning project affects the livelihoods of the herders, could better illustrate the PRP's impact on the herders.

3. Materials and Methods

3.1. Overview of the Study Area and PRP

Maduo County is located in the northeastern part of the Qinghai–Tibet Plateau, at the northern foot of the Bayan Har Mountain, between 96°50′ E–99°20′ E, 33°50′ N–35°40′ N. The administrative division is subordinate to the Guoluo Tibetan Autonomous Prefecture in Qinghai Province, and is the core component of Sanjiangyuan National Park. The average altitude is above 4200 m, and the terrain is relatively flat, with many low mountains, wide valleys and lake basins. It belongs to an alpine grassland climate, with a cold season for more than 10 months in a year. The annual average temperature is −4.1 °C, and the annual precipitation is about 303.9 mm, mostly from May to June. There are dense rivers and numerous lakes, accounting for 43% of the total runoff of the Yellow River Basin. The county has a total area of about 25,253 km^2, and the usable grassland area accounts for about 90% of the county's total area.

Maduo County currently administers 4 townships, namely Huanghe Town, Zalenghu Town, Huashixia Town and Machali Town. It has the smallest prefectural population in Qinhai Province, with a total population of 144,490, of which the Tibetan population accounts for more than 90%. Animal husbandry is the local pillar industry. In the 1980s, herders' per capita net income in Maduo County once ranked first in the country [40]. However, due to the severe deterioration of the ecological environment in the 1990s, the development of local animal husbandry was hindered. After the establishment of the reserve and implementation of ecological restoration projects since 2000, local animal husbandry has been developing slowly due to strict protection restrictions, and it became a national key county for poverty alleviation in 2011. At present, the income of local herders is mainly based on policy subsidies, animal husbandry, and other jobs. In 2018, the annual per capita income of residents was CNY 7401, of which policy subsidies accounted for about 57% and animal husbandry income accounted for about 40%.

The PRP includes a series of specific measures. In order to reduce the livestock number on grassland, complete grazing prohibition and forage–livestock balance, as well as some supporting policies, such as ecological compensation and ecological immigration, fence construction, livestock barn construction, grass storage cellar construction, etc. are included; in order to promote the restoration of grassland vegetation, reseeding grass seed, artificial grassland construction, restoration of black beach and sandy land, prevention and control of poisonous grass and plateau pika are all included. In 2003, the government of Qinghai Province began to implement the PRP on 94.5% of the province' s grasslands, covering an area of 3.45×10^5 km^2. Due to its geographical location at the source of the Yellow River, Maduo County has a wide grazing prohibition area. In 2019, the grazing prohibition area was about 1.67 km^2, and the forage-livestock balance area was 0.58 km^2, accounting for 74.22% and 25.78% of the total usable grassland area, respectively. There have been two large-scale migrations. The first time was the ecological migration from 2004 to 2007, which relocated about 2022 herders to collective villages, accounting for about a quarter of the population at that year. The second time was the herders' relocation for poverty alleviation from 2016, where new houses were built around the county town for 4473 herders, covering more than one third of the population in that year.

At present, the compensation standard for grazing prohibition in Maduo County is CNY 4.1/mu (one mu is approximately 666.67 m^2), and that for forage-livestock is CNY 2.5/mu. Due to the change of family population over the years, the per capita grassland area has become no longer balanced. Therefore, the local government has coordinated various subsidies from the superior government according to the actual situation. The existing subsidy system includes grassland subsidies (approximately CNY 9000 per person per year), ecological inspector subsidies (approximately CNY 21,600 per year), and groups in difficulties (people under the age of 16 and over the age of 55, CNY 5600 per year), subsistence allowance (about CNY 3600 a year), pensions of CNY 200 a month, etc., and the subsidies received by each family vary according to their family situation.

3.2. Data Source

Using the participatory rural evaluation method, the research group travelled to Maduo County twice in August 2020 and July 2021. The first time was a preinvestigation, mainly to understand the implementation of PRP and other policy arrangements in Maduo County. For the second time, a large-scale survey of herders was officially carried out. Due to language barriers, we designed a structured questionnaire and hired and trained four local college students to assist in filling out the questionnaire (see blank informed consent in Appendix A). Using the stratified random sampling method, we successively investigated herders in Guoluo New Village, Tongde County, Hainan Tibetan Autonomous Prefecture (an immigrant village founded in 2006), Heyuan New Village, Maqin County, Guoluo Tibetan Autonomous Prefecture (an immigrant village founded in 2004), Machali

Village, Maduo County, Guoluo Tibetan Autonomous Prefecture (an immigrant village founded in 2007). Although the herders settle in other administrative districts, Guoluo New Village is under the management of Huanghe Township, and Heyuan New Village is under the management of Zalinghu Township. Due to the earthquake on 22 May 2021, many herders in different towns were intensively settled to receive government assistance, so we were able to investigate at various resettlement sites. Figure 1 showed several major survey points, and random surveys scattered over grasslands were not identified. Finally, a total of 266 respondents filled out the questionnaire. Among them, the herders in Huanghe Town, Zhalinghu Town, Machali Town, and Huashixia Town accounted for 36.09%, 31.20%, 20.30%, and 12.41% respectively. In total, 28.57% were intercity immigrants, 25.56% were intercounty immigrants, and 45.86% were local settlers.

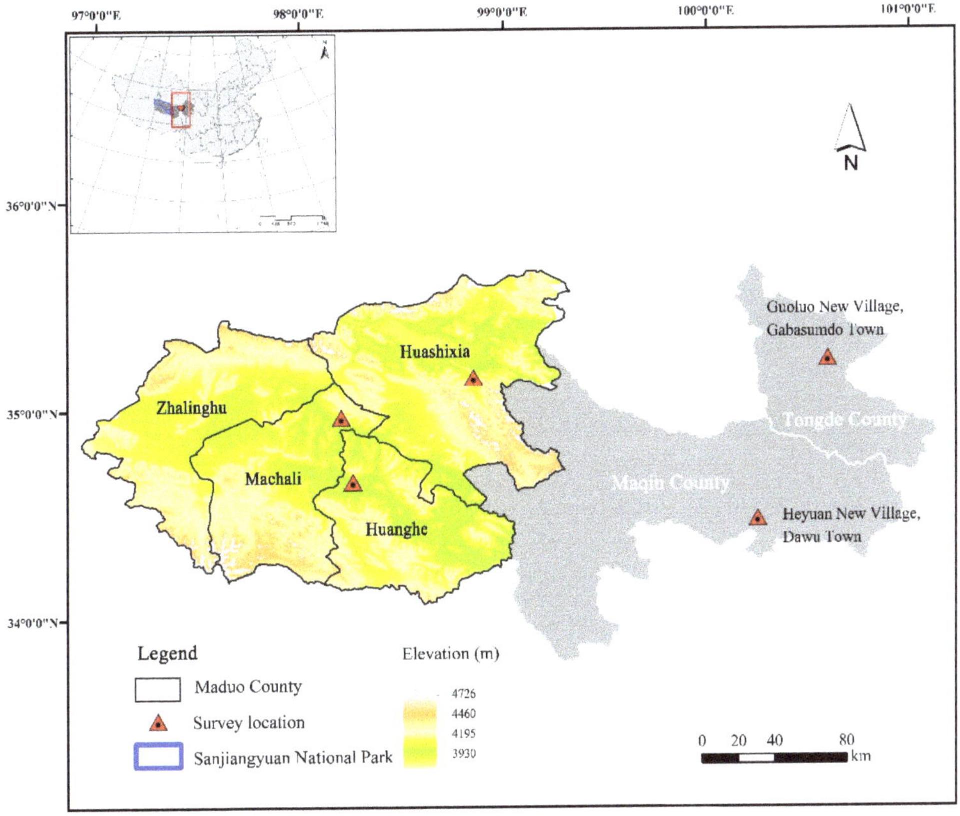

Figure 1. Location of research area.

3.3. Research Methods

In recent decades, self-reported life satisfaction has become an available indicator to measure quality of life [9]. Although a multi-item scale has higher validity and reliability [5,41], the one-dimensional satisfaction scale has also been widely used in measuring overall life satisfaction or satisfaction in one field [10,11,27]. Taking life satisfaction as the dependent variable, a single scale of "are you satisfied with your current life?" was designed to measure the life satisfaction of herders in Maduo County from very dissatisfied to very satisfied according to the division of Likert's five-level scale. In addition, the satisfaction of herders in other main areas of life was investigated, including housing, medical treatment, education, work, income, consumption, leisure, neighborhood, etc.

Multiple linear regression could be used to analyze the relationship between the indicators in the subjective evaluation scale [11]. In our study, the model was used to analyze the impact of satisfaction in different aspects on overall life satisfaction, in order to determine which aspects of life have a significant association with life satisfaction, and the degree of their respective effects. Stepwise regression was adopted to directly screen out the key life aspects that have significant impacts on overall life satisfaction. The principle of this method is to recalculate significance of the substituted independent variables after introducing each new independent variable, and judge whether it is necessary to continue to remain them in the equation. After repeated introduction and elimination of independent variables until no new significant variables could be introduced or removed, an optimal explanatory model was finally obtained, and the independent variables remaining in the model were all ones that have a significant effect on the dependent variable.

In order to explore the influencing factors of herders' life satisfaction under the implementation of PRP, three-dimensional influencing factors were listed according to the actual situation of herders in Maduo County and previous research (see Table 2). The first dimension was the demographic characteristics of herders and the socioeconomic characteristics of their families, mainly involving the respondents' gender, age, education, occupation, income, etc.; second, we paid attention to the impact of the natural and social environments on herders' life satisfaction. In terms of natural environment, herders' evaluation of natural environment quality was selected as the representative indicator of the impact of the natural environment on their lives; in terms of social environment, we focused on the support for herders' livelihood in three aspects—herders' ability, social relations, and government assistance—and selected relative living standard, difficulty in borrowing, market convenience, and government service satisfaction as representative indicators. The difficulty in borrowing reflects the possibility of herders obtaining external assistance in case of economic difficulties, and the market convenience reflects the difficulty of herders in livestock trading, working, and self-employment. Third, we paid special attention to whether the direct impact of the PRP project on herders would affect herders' life satisfaction, which mainly involved three aspects, those being immigration, income, and grassland restoration, with settlement site, herders' perception of income impact, subsidy level, and revegetation effect as the specific indicators. The assignment of all indicators is shown in Table 2. Based on previous research, we hypothesized that the availability of social support may be positively correlated with life satisfaction, because it would meet herders' needs of safety. The project of PRP may have a negative impact on the life satisfaction of herders, which may be due to the reduction of animal husbandry income after grazing prohibition and the maladjustment after the change of their traditional lifestyle.

Since the dependent variable and most of independent variables are subjectively ordinal multicategorical variables but not precise quantitative variables, traditional linear regression could not accurately reflect the actual meaning of those variables. Otherwise, in the logistic regression model, dummy variables should be set for multilevel independent variables, respectively. In our study, the model results may struggle to accurately reflect the actual situation due to too many categories of independent variables and the small number of samples after setting dummy variables. Finally, we found that it is more appropriate to select the optimal scale regression model.

Optimal scaling regression is a new program after SPSS11.0 developed by DTSS research group of Leiden University in the Netherlands (The version we use is SPSS 27.0, released by IBM in Armonk, NY, USA). This method breaks through the limitation of categorical variables on model selection. In the analysis, according to the influence of each level of each independent variable on the dependent variable, a nonlinear transformation method is used to convert the values of each level of the original categorical variables on the premise that the linear relations of transformed variables are maintained, and the converted values are used to replace the original values for model construction. After repeated iterations, the best model is finally constructed. The difference between the converted values approximately indicates the difference in the influence degree of different levels of independent variables on the dependent variable. A large difference in two values indicates a large difference in the influence between corresponding levels. In this way, the differences between various levels can be quickly found and the possible irrationality of the original assignment can be avoided. The model structure is similar to the general linear regression, as shown below. x_i represents the independent variable, ε_i represents the random error.

$$y_i = \beta_0 + \beta_i x_i + \varepsilon_i \tag{1}$$

Table 2. List of independent variables.

Dimension	Variable Name	Variable Assignment
Personal characteristics	Gender	Male = 1; Female = 2
	Age	Quantitative variable
	Education	Illiterate = 1; Primary school = 2; Junior high school = 3; High school = 4; College or above = 5
	Occupation	Unemployed = 1; Independent employment = 2; Employed by government = 3
	Willingness to change	Don't want to change at all = 1; Obedience to government arrangement = 2; Depending on the situation = 3; Want to change = 4; Eager to change = 5
	Health	Incapacity = 1; Poor = 2; Medium = 3; Good = 4; Excellent = 5
Family characteristics	Livestock retention	No= 1; Yes= 2
	Annual income	Less than CNY 30 thousand = 1; CNY 30–50 thousand = 2; CNY 50–100 thousand = 3; Over CNY 100 thousand = 4
Perception of the environment	Natural environment	Very poor = 1; Poor = 2; Fair = 3; Good = 4; Very good = 5
	Relative living standard	Lower = 1; Lower-middle = 2; Middle = 3; Upper-middle = 4; Upper = 5
	Difficulty in borrowing	Very difficult = 1; Relatively difficult = 2; Depending on the situation = 3; Relatively easy = 4; Very easy = 5
	Market convenience	Very inconvenient = 1; Inconvenient = 2; Fair = 3; Convenient = 4; Very convenient = 5
	Government services	Very dissatisfied = 1; Dissatisfied = 2; Neither satisfied nor dissatisfied= 3; Satisfied = 4; Very satisfied = 5
Impact of PRP	Immigration	Trans-city migration = 1; Trans-county migration = 2; Local residence = 3
	Income change	Decrease a lot = 1; Decrease a little = 2; No change = 3; Increased a little = 4; Increase a lot = 5
	Subsidy level	Very few = 1; few= 2; Appropriate = 3; Many = 4; Too many = 5
	Grassland restoration	Very poor = 1; Poor = 2; Fair = 3; Good = 4; Very good = 5

4. Results

4.1. Descriptive Analysis

The proportion of some basic characteristics of the respondents was presented in Table 3. Of all respondents, men and women accounted for 55% and 45%, respectively, 81.95% were between 20 and 60 years old. In total, 76.69% of respondents had never attended school, indicating that the heads of households in this area generally have a low level of education, which may have a negative impact on family development. The households with a large number of livestock (more than 10) accounted for only 8.64% of the surveyed samples, and most households did not have livestock. One reason for this result may be that many herders have abandoned livestock breeding or converted their livestock into cooperative as shares after the large-scale grazing prohibition in Maduo County; the other one may be that the time of the survey coincided with herders who had a large number of livestock moving to a remote summer pasture, making it difficult to find more samples. In terms of employment, only 31.95% of households had their own income channels, such as grazing, formal work, part-time jobs, self-employed entrepreneurs. Nearly 50% of herders were only engaged in positions provided by the government, such as ecological inspector and cleaner. In addition, nearly 20% of households had no work. In terms of household income, the proportions from high to low were CNY 30 to 50 thousand (43.99%), CNY 50 to 100 thousand (32.71%), less than CNY 30 thousand (16.92%), and greater than CNY 100 thousand (5.26%). A total of 62.93% of herders believed that expenditure was greater than income, and food and medical care are the main expenses. Only 9.05% of households felt income greater than expenditure. It could be seen that the economic base of herders in Maduo County is weak. The livelihood of most households relies on cash subsidies and job provided by government, and their consumption is mainly concentrated on daily necessities, and there is a lack of reserves to cope with risks. This section may be divided by subheadings. It should provide a concise and precise description of the experimental results, their interpretation, as well as the experimental conclusions that can be drawn.

Table 3. Basic characteristics of herders' livelihood.

Category	Types	Proportion	Category	Types	Proportion
Gender	Male	55.26%	Livestock retention	No	91.35%
	Female	44.74%		Yes	8.65%
Age	<20	4.14%	Occupation	Unemployed	19.55%
	20–40	36.84%		Independent employment	31.95%
	40–60	45.11%		Employed by government	48.50%
	>60	13.91%	Annual income	Less than CNY 30 thousand	16.92%
Education	Illiterate	76.69%		CNY 30~50 thousand	43.98%
	Primary school	10.90%		CNY 50~100 thousand	32.71%
	Junior high school	5.64%		Over CNY 100 thousand	5.26%
			Balanced budget	Expenditure exceeds income	62.93%
	High school	3.76%		Expenditure equals income	28.02%
	College or above	3.01%		Income exceeds expenditure	9.05%

Figure 2 showed the herders' subjective evaluation of each indicator in the two dimensions of the external environment and the impacts of PRP. According to herders' emotional attitude, the original 5-point values were divided into three categories: negative, neutral, and positive. The broken line represented the mean score of each indicator. As shown in Figure 2, the average scores of the difficulty in borrowing, relative living standard, change in income and subsidy level were less than 3, indicating that the herders' comprehensive evaluation of these aspects was relatively negative. Specifically, the negative evaluation of difficulty in borrowing and change in income accounted for the largest proportion, and the neutral evaluation of relative living standard and subsidy level accounted for the largest proportion. The scores of natural environment, market convenience, satisfaction with government services, and ecological effects were greater than 3, and the proportions of herders holding a positive evaluation of these aspects were much higher than that of negative and neutral evaluation. The above results show that the majority of herders were satisfied with the local natural conditions, even if the climate indicators show it was very bad. There was little difference in the perception of rich and poor among local herders, and more than half of herders believed that their living conditions were at a medium level. Many herders had some difficulties in borrowing money or getting a loan, which may affect their ability to resist risks. The local market environment was good, and it was not very difficult for herders to participate in market activities. The PRP had an obvious negative impact on the income of herders, but the subsidies had well offset some of the losses. Herders had a very positive evaluation of the government services and the effect of PRP on grassland restoration.

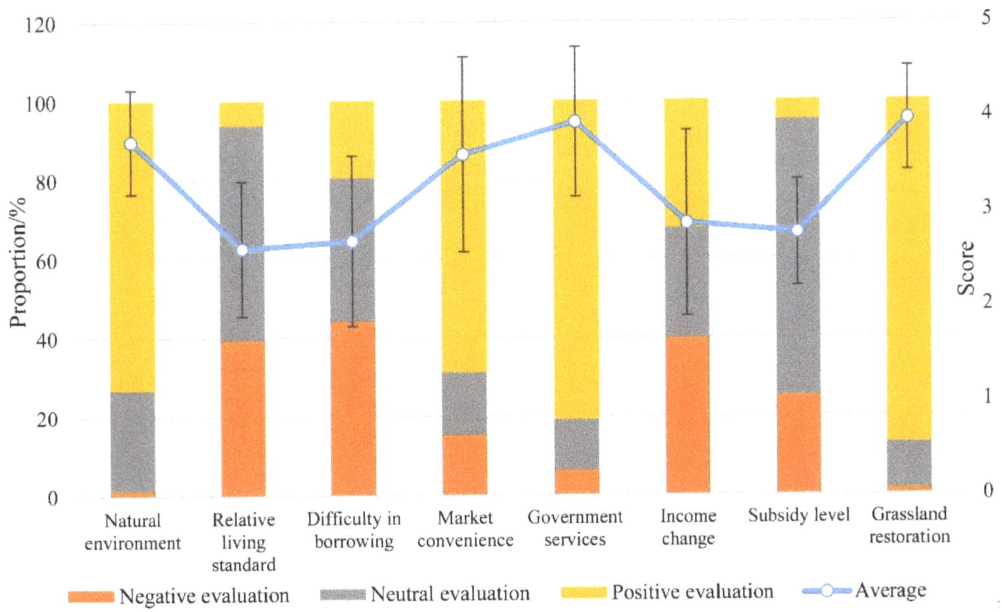

Figure 2. Herders' perception of environment and PRP's impact.

4.2. Descriptive Analysis of Herders' Satisfaction

The proportion of respondents' satisfaction scores of 1–5 in all areas of life was presented in Figure 3, and the black triangle represented the average of satisfaction scores. It can be seen that herders had high satisfaction about several fields of life and overall life satisfaction, with all average values greater than 3. Among them, the satisfaction about neighborhood relations, leisure life, and work was the top three, and the overall life satisfaction was less than that of any single field. Since life satisfaction is a comprehensive assessment of personal status, high satisfaction in a certain field may be offset by low satisfaction in a certain field [42]; therefore, the overall low life satisfaction of herders indicates that some fields with low satisfaction may not be paid attention to.

The Pearson correlation matrix analysis of each variable shows that the overall satisfaction was significantly correlated with other fields' satisfaction at the level of 0.01 or 0.05 except neighborhood relationship. In order to understand the contribution of satisfaction in various fields to overall life satisfaction, SPSS software was used to carry out multiple liner regression, and the step-by-step method was used to screen variables and construct the model. In the end, only consumption satisfaction and leisure satisfaction were selected into the optimal model, with coefficients of 0.273, 0.253 respectively, indicating that compared with other variables, these two variables had a greater impact on overall life satisfaction. However, R^2 of the model was 0.165, meaning that the two variables had a relatively low interpretation of overall life satisfaction.

4.3. Analysis of Determinants of Life Satisfaction

Because both independent variables and dependent variables were ordered classification variables, χ^2 test was used to judge whether each independent variable was associated with life satisfaction. Because the theoretical frequency of some levels in different independent variables were less than 1, we referred to the results of Fisher's exact test. The results (shown in Table 4) show that except for occupation, health status, whether grazing or not, and annual family income, other variables were significant at the level of 0.05 or 0.01, indicating that these variables were related to life satisfaction. However, only according to the results of χ^2 test, the joint impact of multiple independent variables on life satisfaction and the contribution of each variable cannot be judged, and multivariable regression analysis needs to be further conducted.

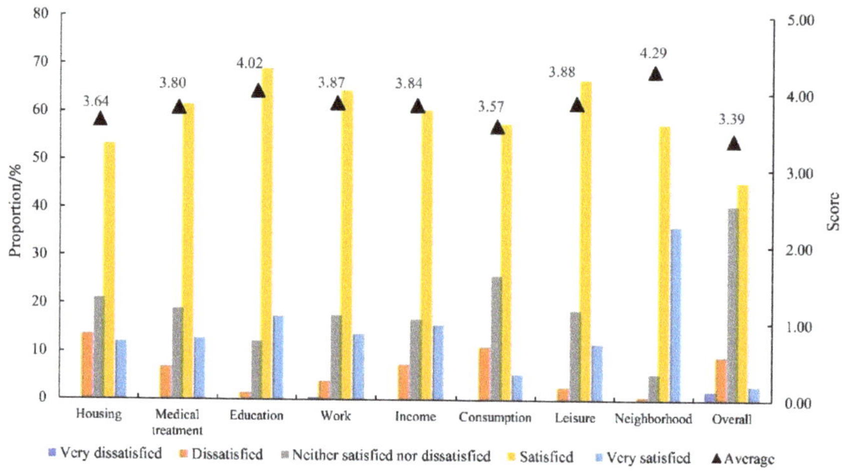

Figure 3. Herders' subjective evaluation of life satisfaction.

Table 4. Results of χ^2 test.

Variable Name	Fisher's Exact Test	p Value	Variable Name	Fisher's Exact Test	p Value
Personal characteristics			*Perception of the environment*		
Gender	9.850	0.038	Natural environment	44.518	0.000
Education	29.857	0.007	Difficulty in borrowing	28.903	0.009
Occupation	12.215	0.104	Relative living standard	48.559	0.000
Willingness to change	28.163	0.012	Market convenience	41.379	0.000
Family characteristics			Government services	21.288	0.000
Health	20.858	0.093	*Impact of PRP*		
Livestock retention	6.759	0.110	Immigration	23.271	0.001
Annual income	11.628	0.401	Income change	46.797	0.000
			Subsidy level	51.379	0.000
			Grassland restoration	26.559	0.003

The optimal scale regression of different dimensional independent variables to life satisfaction was carried out separately. Table 5 showed the results of the four models, including the coefficients and significance of independent variables, the relative importance of variables, the significance of the models and R^2. The characteristic of optimal scale model is to convert the original values and construct the model with the converted values to achieve the optimal fitting effect. The influence difference of different levels of an independent variable on the dependent variable could be expressed by the product of the difference between two values and the coefficient of corresponding variable. In order to judge the difference in the impact of different levels of independent variables on life satisfaction, the converted values of independent variables with significant coefficients at the level of 0.1 were drawn in Figure 4 according to the results in Table 5.

Table 5. Results of optimal scale regression.

Variable Name	Model 1 Coefficient	Model 1 Importance	Model 2 Coefficient	Model 2 Importance	Model 3 Coefficient	Model 3 Importance	Model 4 Coefficient	Model 4 Importance
Personal characteristics								
Gender	0.079	0.044			0.002	0.000	0.033	0.000
Age	0.009	0.001			0.024	0.003	0.059	0.005
Education	0.202 ***	0.335			0.083	0.006	0.078	0.006
Occupation	0.140 **	0.104			0.107 **	0.033	0.092 *	0.012
Willingness to change	−0.214 **	0.367			−0.150	0.064	−0.144	0.048
Family characteristics								
Health	0.026	0.005			−0.018	−0.001	−0.023	−0.004
Livestock retention	0.046	0.021			0.098 *	0.038	0.079 *	−0.023
Annual income	0.120	0.123			0.099	0.040	0.067	0.022
Perception of the environment								
Natural environment			0.139	0.076	0.121	0.083	0.147	0.074
Relative living standard			0.240 **	0.175	0.187 ***	0.158	0.146 **	0.084
Difficulty in borrowing			0.139 ***	0.077	0.145 **	0.086	0.148 ***	0.060
Market convenience			−0.226 ***	0.126	−0.289 ***	0.181	−0.279 ***	0.143
Government services			0.266 **	0.200	0.346 ***	0.310	0.284 ***	0.183
Impact of PRP								
Immigration			0.103 **	0.035			0.134 *	0.045
Income change			0.223	0.130			0.224	0.124
Subsidy level			−0.291	0.154			−0.273	0.159
Grassland restoration			0.069	0.027			0.047	0.016
F test	2.191		7.995		5.175		5.646	
p value	0.005		0.000		0.000		0.000	
R^2	0.133		0.432		0.363		0.465	
Adjusted R^2	0.072		0.378		0.293		0.383	

*, $p < 0.1$; **, $p < 0.05$; ***, $p < 0.01$.

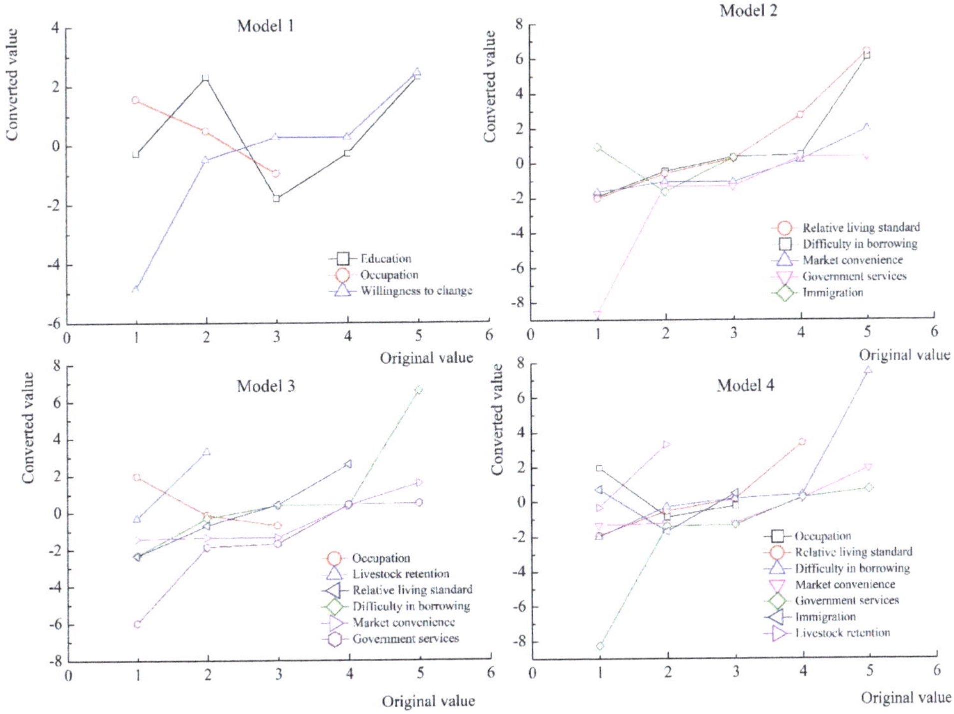

Figure 4. Numerical conversion of significant independent variables.

The results showed that the *p* values of the four models were less than 0.01, indicating that all models were statistically significant. The tolerance of all independent variables was greater than 0.1, indicating that there was no multicollinearity between independent variables. Model 4 included all variables, with an R^2 of 0.465, indicating that the equation fitted according to selected independent variables can explain 46.5% of the variance of life satisfaction. Model 1 only included variables of individual and family characteristics, with an R^2 of 0.133. Model 2 included variables of herders' perception of environment and the impact of PRP, for which R^2 was 0.432. In Model 1, three variables (education, occupation, and willingness to change) were significant, but after adding other variables in Model 3 and Model 4, education and willingness to change became insignificant, and the standardized regression coefficient of occupation decreased, although it remained significant. Compared with Model 3, after adding the variables of the impact of PRP, the R^2 of Model 4 only increased by 0.102, and only the variable of immigration was significant. The above results showed that in the four dimensions, the herders' perception of the external environment had the greatest impact on life satisfaction. To improve herders' life satisfaction, the social environment in which the herders live should be optimized.

4.3.1. Personal and Family Characteristics

Some studies have confirmed that the relationship between demographic variables and life satisfaction was weak [41]. In our study, the significance of independent variables of individual and family characteristics changed in different models, and the explanation of these characteristics to life satisfaction decreased with the addition of more variables. In Model 1, education, occupation and willingness to change had a significant impact on life satisfaction. For every unit of education improvement, life satisfaction will increase by 0.202 units. Combined with the conversion value, herders with primary school education and college education both had higher life satisfaction. After numerical conversion, the coefficient of occupation was 0.140, indicating that unemployed herders had the highest life satisfaction, followed by self-employment, public welfare position. According to the actual situation in Maduo County, the high satisfaction of unemployed herders may be due to the fact that their livelihoods can be maintained by subsidies and a lot of leisure time could be spent even they are incompetent for work or do not want to be employed. Pubic welfare position mainly refers to ecological managers. The reason for their low satisfaction may be that, on the one hand, the selected ecological managers inherently have relatively difficult family conditions. On the other hand, although there is a salary of about CNY 20 thousand a year, the work of frequent patrols is relatively heavy, especially for these managers who live in remote settlements. There was a negative correlation between willingness to change and life satisfaction. Life satisfaction decreased by 0.214 units when willingness to change increased by one unit, which suggests that dissatisfaction with reality may stimulate herders to make some changes. In Model 3 and Model 4, livestock retention became new positive significant factors at the level of 0.1, showing that herders with livestock were more satisfied with their lives.

4.3.2. Perception of Natural and Social Environment

In the dimension of environmental perception, the natural environment had no significant impact on the herders' life satisfaction, while all variables about social environment had a significant impact in each model and the coefficients of all variables were positive except market convenience. The higher the relative living standard, the higher the life satisfaction. Combined with the conversion value in model 3, the life satisfaction of herders in the middle class and upper-middle class increased by 0.52 and 0.92 units, respectively, compared with the lower class. With the reduction of the difficulty in borrowing, the life satisfaction of herders gradually improved. The life satisfaction of herders who felt very easy was 1.29 units higher than that of herders who felt very difficult. Satisfaction with government services was also positively correlated with life satisfaction, indicating that improving the service level of government agencies could be an effective way to improve herders' life satisfaction. In term of market convenience, compared with herders who think it was very difficult to engage in market activities, the life satisfaction of herders who feel it was very convenient was reduced by 0.87 units. This may be because herders who think it was very convenient to engage in market activities might engage in short-term employment. Although such jobs can be found, their contribution in improving herders' life quality may be limited. Herders who think it was difficult to participate in market activities may be actively carrying out self-employed activities and experiencing more practical difficulties, but their personal abilities, family conditions and risk resistance capabilities generally stronger than surrounding families.

4.3.3. The Impacts of PRP

In Model 2 and Model 4, only the variable of immigration was significant at the level of 0.05 and 0.1 respectively. Combined with the conversion value in Model 3, the life satisfaction of different immigrant groups in descending order was transcity immigrants, local residence, and transcounty immigrants. According to our investigation, the transcity settlement, Guoluo New Village, is far away from nearby towns, and the surrounding land belongs to other towns and cannot be used. Most livestock greenhouses built at the beginning have been abandoned. Compared with other places, the livelihood development of the herders in this village should be relatively disadvantageous. However, the results show that the herders in Guoluo New Village were highly satisfied, which may be because after the earthquake, most of the residents of Guoluo New Village returned to live in tents around the Maduo county, and the proper resettlement of the government may promote the herders to make a higher life satisfaction evaluation at that time.

Many studies have suggested that the positive impact of the policies on income and the improvement of subsidy level can promote the improvement of life satisfaction [22]. From our results, compared with other variables, the effects of the two were not significant, the importance of income change, subsidy level accounted for more than 10%, indicating that these two variables were more important in explaining life satisfaction than other independent variables and the greater the positive impact of income, the higher herders' life satisfaction. However, too many subsidies may have no positive effect on improving herders' life satisfaction. It could be observed that the direct impact of PRP on herders' life satisfaction might be not obvious, but the PRP could indirectly affect herders' life satisfaction through a number of measures, such as the performance of the government during the implementation of the program, the relative income level of herders after subsidy, new livelihood and social relations after resettlement, etc.

5. Conclusions and Discussion

This paper analyzed herders' life satisfaction and its influencing factors under the PRP in Maduo County, the source of the Yellow River. Our results suggested that the life satisfaction of herders in Maduo County was relatively high, which was largely due to the government's strong subsidies and the overall improvement of people's livelihood under the background of poverty alleviation and Sanjiangyuan National Park construction in recent years. The positive impact of PRP on the lives of herders in Maduo County has increased in recent years. At present, the subsidy level and the impact of PRP on income were no longer significant factors influencing the satisfaction of herders, although they were also very important. The gap between the rich and the poor, the level of assistance available, and the quality of government services had a greater impact on the satisfaction of herdsmen.

The difference in socioeconomic situation and government assistance in different regions will result in different impacts of the PRP on herders. The Sanjiangyuan area, especially the area where the Sanjiangyuan National Park is located, has its own characteristics, which distinguish it from other implementation areas of PRP. Firstly, the grazing prohibition area is much larger than the forage–livestock balance area, and there is no available cultivated land for herders to plant forage to feeding livestock. After the grazing prohibition, many herders moved to cities and towns and only a few herders still stayed in the grazing prohibition area. Therefore, many herders have experienced a rapid change in their lifestyle in a short time and can no longer continue to engage in animal husbandry. Secondly, after the migration, geographical location, employment environment and language barriers make it difficult for local herders to move to other areas for reemployment such as herders in Gansu, Ningxia, and Inner Mongolia [43]. In addition, with the change of lifestyle, herders need to spend much money on daily necessities imported from outside, such as food and fuel, and travel to more developed areas for medical treatment and education. Therefore, the immigrants are more dependent on government subsidies than other areas. Finally, although the natural and social environment is very unfavorable, the state's high subsidies and assistance to this region are also different from other regions. Now, the main source of income of herders, especially the migrants, was government cash subsidies and subsidies for public-service jobs. However, the subsidy system may not be feasible in other areas due to their lower ecological status, huge population size and availability of alternative livelihoods.

The uniqueness of each region means that the policies should be more targeted and flexible to promote the well-being of residents. The government usually focuses on promoting income growth and optimizing infrastructure to improving people's well-being. The efforts in Maduo County by optimizing infrastructure such as medical treatment, education, transportation, and electricity have received positive feedback, which is reflected in relatively high satisfaction of herders in the main aspects of life, although the objective conditions may be far inferior to other developed areas. Increasing subsidies has always been regarded as a way to ensure the well-being of local people in the implementation area of PRP [44]. However, the improvement of absolute income and life satisfaction may not keep pace [45,46], which was also confirmed by our research. In Maduo County, nearly 70% of herders think the subsidy was appropriate at present. In terms of family economic conditions, the impact of relative living standards is more important than that of absolute income and income changed by PRP. In addition, the possibility of obtaining support from relatives, friends, and the government would significantly affect herders' life satisfaction. In terms of the importance of all areas of life to overall life satisfaction, consumption and leisure life had a greater impact, which is consistent with some research results [47]. These results mean that the psychological needs of local herders are more important than material needs in the overall life satisfaction. Therefore, the optimization of PRP should start by controlling the gap between the rich and the poor, optimizing social assistance channels, and improving government service level.

In addition, under the current subsidy system, herders' livelihoods have been basically guaranteed. Herders who are unemployed or do not want to change the current situation have higher life satisfaction, but the high life satisfaction maintained by subsidies may weaken herders' own development potential and ability to deal with risks in the future. Herders themselves have the motivation to make some change when they feel dissatisfied with life, but whether they can change in the end is affected by many factors. In terms of policy design, on the one hand, the original subsidies should not be reduced at the beginning of active development, which may lead to herders' fear of risks and complacency with the status quo. On the other hand, education and vocational training for herders and their next generation should be strengthened to enable them to own more opportunities for self-development, and to reduce their dependence on subsidies.

Although this article could contribute to understanding the life satisfaction of local herders, it has some limitations for policy reference. The first is that the high self-rated life satisfaction of herdsmen may not mean that the local development level and herdsmen's living standard are consistent with those in more developed areas, but may be due to the narrow reference standard of herders [48]. The second is that 53.5% of the variance in life satisfaction remained unexplained. Since there is a satisfaction baseline based on people's own characteristics [49], for policy makers, the most concern should be placed on the part that can promote life satisfaction through policy design. Our research has found the differences in life satisfaction of different types of herders, so the differences in the demands of families with different conditions need to be identified and more effective assistance are needed in the future.

There are still many issues worthy of further study. First, due to the numerous factors that influence life satisfaction and the uniqueness of Tibetan culture, future research can explore its impact on life satisfaction from the perspectives of herders' own personalities, comparative standards, and religious influences, etc. Second, in the future multidimensional scales with higher reliability could be used to further explore the relationship between life satisfaction and satisfaction in various fields, as well as the influencing factors of satisfaction in various fields, so as to find specific ways to improve local people's well-being. Finally, due to the particularity of Maduo County, in the future, more examples should be studied to compare the differences in the impact of policies.

Author Contributions: Conceptualization, C.X., L.Z. and Y.W.; methodology, C.X. and Y.W.; data collection, C.X., L.Z., Y.W. and X.P.; writing original draft and preparation, C.X.; data analysis, C.X. and Y.W. All authors have read and agreed to the published version of the manuscript.

Funding: This research was funded by the Second Tibetan Plateau Scientific Expedition and Research Program (2019QZKK0404), The Strategic Priority Research Program of the Chinese Academy of Sciences (XDA20020401) and Joint Grant from Chinese Academy of Sciences–People's Government of Qinghai Province on Sanjiangyuan National Park (LHZX-2020-09).

Institutional Review Board Statement: Ethical review and approval were waived for this study due to this study was not for medical purposes.

Informed Consent Statement: Informed consent was obtained from all subjects involved in the study.

Conflicts of Interest: The authors declare no conflict of interest.

Appendix A. Blank Informed Consent

Questionnaire for herders of performance evaluation of the Pastureland Rehabilitation Program in Sanjiangyuan area

Hello! We are researchers from the Institutes of Science and Development, Chinese Academy of Sciences. We are here to conduct a questionnaire survey on the impact of the Pastureland Rehabilitation Program on the herders' livelihood. This survey is completely anonymous, and we will not disclose any information about you. Please feel free to answer. Thank you very much for your support for our work.

References

1. McMahon, D.M. The pursuit of happiness in history. In *The Science of Subjective Well-Being*; Eid, M., Larsen, R.J., Eds.; Guilford: New York, NY, USA, 2008; pp. 80–93.
2. Salovey, P.; Mayer, J.D. Emotional intelligence. *Imagin. Cogn. Personal.* **1990**, *9*, 185–211. [CrossRef]
3. Bailey, T.C.; Eng, W.; Frisch, M.B.; Snyder, C.R. Hope and optimism as related to life satisfaction. *J. Posit. Psychol.* **2007**, *2*, 168–175. [CrossRef]
4. Diener, E.; Diner, M. Cross-Cultural Correlates of Life Satisfaction and Self-Esteem. *J. Personal. Soc. Psychol.* **1995**, *68*, 653–663. [CrossRef]
5. Pavot, W.; Diener, E. Review of the Satisfaction with Life Scale. *Psychol. Assess.* **1993**, *5*, 164–172. [CrossRef]
6. Diener, E.; Inglehart, R.F.; Tay, L. Theory and Validity of Life Satisfaction Scales. *Soc. Indic. Res.* **2012**, *112*, 497–527. [CrossRef]
7. Frey, B.S.; Stutzer, A. What can Economists Learn from Happiness Research? *J. Econ. Lit.* **2002**, *40*, 402–435. [CrossRef]
8. Londoño, C.O.; Mesa, D.G.; Cardona-Sosa, L.; Toro, C.G. Happiness and victimization in Latin America. *J. Happiness Stud.* **2019**, *20*, 935–954. [CrossRef]
9. Alexandrova, A. Subjective well-being and Kahneman's 'objective happiness'. *J. Happiness Stud.* **2005**, *6*, 301–324. [CrossRef]
10. Blanchflower, D.G.; Oswald, A.J. Well-being over time in Britain and the USA. *J. Public Econ.* **2004**, *88*, 1359–1386. [CrossRef]
11. Appleton, S.; Song, L. Life Satisfaction in Urban China: Components and Determinants. *World Dev.* **2008**, *36*, 2325–2340. [CrossRef]
12. Arrindell, W.A.; Heesink, J.; Feij, J.A. The satisfaction with life scale (SWLS): Appraisal with 1700 healthy young adults in The Netherlands. *Personal. Individ. Differ.* **1999**, *26*, 815–826. [CrossRef]
13. Ortega-Gil, M.; Cortés-Sierra, G.; ElHichou-Ahmed, C. The Effect of Environmental Degradation, Climate Change, and the European Green Deal Tools on Life Satisfaction. *Energies* **2021**, *14*, 5839. [CrossRef]
14. Mirehie, M.; Sato, S.; Krohn, B. Participation in Active Sport Tourism and Life Satisfaction: Comparing Golf, Snowboarding, and Long-Distance Running. *Sustainability* **2021**, *13*, 10316. [CrossRef]

15. Berlemann, M.; Eurich, M. Natural hazard risk and life satisfaction–Empirical evidence for hurricanes. *Ecol. Econ.* **2021**, *190*, 107194. [CrossRef]
16. Easterlin, R.A.; Morgan, R.; Switek, M.; Wang, F. From the cover: China's life satisfaction, 1990–2010. *Proc. Natl. Acad. Sci. USA* **2012**, *109*, 9775–9780. [CrossRef] [PubMed]
17. Fan, J.; Zhong, H.; Harris, W.; Yu, G.; Wang, S.; Hu, Z.; Yue, Y. Carbon storage in the grasslands of China based on field measurements of above- and below-ground biomass. *Clim. Chang.* **2008**, *86*, 375–396. [CrossRef]
18. Hu, Y.N.; Huang, J.K.; Hou, L.L. Impacts of the Grassland Ecological Compensation Policy on Household Livestock Production in China: An Empirical Study in Inner Mongolia. *Ecol. Econ.* **2019**, *161*, 248–256. [CrossRef]
19. Hou, C.; Zhou, L.; Wen, Y.; Chen, Y. Farmers' adaptability to the policy of ecological protection in China-A case study in Yanchi County, China. *Soc. Sci. J.* **2018**, *55*, 404–412. [CrossRef]
20. Byrne, A.T.; Hadrich, J.C.; Robinson, B.E.; Han, G.D. A factor-income approach to estimating grassland protection subsidy payments to livestock herders in Inner Mongolia, China. *Land Use Policy* **2020**, *91*, 104352. [CrossRef]
21. Li, M.Y.; Zhao, P.J.; Wu, L.B.; Chen, K. Effects of Value Perception, Environmental Regulation and Their Interaction on the Improvement of Herdsmen's Grassland Ecological Policy Satisfaction. *Int. J. Environ. Res. Public Health* **2021**, *18*, 3078. [CrossRef]
22. Wang, L.; Liu, X. Satisfaction with grassland eco-compensation policies for herders: An empirical study on the Gansu Pastoral Area. *Acta Ecol. Sin.* **2017**, *37*, 5798–5806. (In Chinese)
23. Johnson, D.C.S.M. Avowed happiness as an overall assessment of the quality of life. *Soc. Indic. Res.* **1978**, *5*, 475–492. [CrossRef]
24. Diener, E.; Suh, E.M.; Lucas, R.E.; Smith, H.L. Subjective well-being: Three decades of progress. *Psychol. Bull.* **1999**, *125*, 276–302. [CrossRef]
25. Helliwell, J.F. How's life? Combining individual and national variables to explain subjective well-being. *Econ. Model.* **2003**, *20*, 331–360. [CrossRef]
26. Diener, E.; Oishi, S.; Tay, L. Advances in subjective well-being research. *Nat. Hum. Behav.* **2018**, *2*, 253–260. [CrossRef] [PubMed]
27. Diener, E.; Oishi, S.; Lucas, R.E. Subjective well-being: The science of happiness and life satisfaction. In *Oxford Handbook of Positive Psychology*; Snyder, C.R., Lopez, S.J., Edwards, L.M., Marques, S.C., Eds.; Oxford University Press: Oxford, UK, 2009; pp. 187–194.
28. Maslow, A.H. *Motivation and Personality*; Harper & Row: New York, NY, USA, 1970.
29. Michalos, A.C. Multiple discrepancies theory. *Soc. Indic. Res.* **1985**, *16*, 347–413. [CrossRef]
30. Shields, M.A.; Price, S.W.; Wooden, M. Life satisfaction and the economic and social characteristics of neighbourhoods. *J. Popul. Econ.* **2009**, *22*, 421–443. [CrossRef]
31. Yin, Y.T.; Hou, Y.L.; Langford, C.; Bai, H.H.; Hou, X.Y. Herder stocking rate and household income under the Grassland Ecological Protection Award Policy in northern China. *Land Use Policy* **2019**, *82*, 120–129. [CrossRef]
32. Yu, Y.; Wu, Y.; Wang, P.; Zhang, Y.L.; Yang, L.E.; Cheng, X.; Yan, J.Z. Grassland Subsidies Increase the Number of Livestock on the Tibetan Plateau: Why Does the "Payment for Ecosystem Services" Policy Have the Opposite Outcome? *Sustainability* **2021**, *13*, 6208. [CrossRef]
33. Wang, Y.; Han, Y.Q.; Han, Y.J.; Li, W.C. Does the grassland ecological compensation policy improve the herders' breeding technical efficiency in china?-based on the parallel mediation effect model. *PLoS ONE* **2021**, *16*, e0249990. [CrossRef]
34. Liu, M.; Rao, D.; Yang, L.; Min, Q. Subsidy, training or material supply? The impact path of eco-compensation method on farmers' livelihood assets. *J. Environ. Manag.* **2021**, *287*, 112339. [CrossRef] [PubMed]
35. Zhen, L.; Li, F.; Yan, H.M.; Liu, G.H.; Liu, J.Y.; Zhang, H.Y.; Du, B.Z.; Wu, R.Z.; Sun, C.Z.; Wang, C. Herders' willingness to accept versus the public sector's willingness to pay for grassland restoration in the Xilingol League of Inner Mongolia, China. *Environ. Res. Lett.* **2014**, *9*, 045003. [CrossRef]
36. Ding, R.; Shao, L.Q.; Chen, H.B. Curbing overstocking on rangeland through subsidies, rewards, and herders' social capital: Lessons from Qinghai province, China. *J. Rural. Stud.* **2021**, *87*, 361–374. [CrossRef]
37. Wang, W.; Zhou, L.; Yang, G.; Sun, Y.; Chen, Y. Prohibited grazing policy satisfaction and life satisfaction in rural northwest China-a case study in Yanchi County, Ningxia Hui Autonomous Region. *Int. J. Environ. Res. Public Health* **2019**, *16*, 4374. [CrossRef]
38. Li, Y.; Wei, T.; Jin, L. Herdspeople Attitudes Towards Grassland Eco- Compensation Policies in Siziwang Banner, Inner Mongolia. *Resour. Sci.* **2014**, *36*, 2442–2450. (In Chinese)
39. Wang, W.; Yang, G.; Sun, Y.; Chen, Y.; Zhou, L. Linking prohibited grazing policy to farmers' subjective well-being: A case study in Yanchi County, China. *Sustainability* **2019**, *11*, 2180. [CrossRef]
40. Zhang, Z.L.; Xiang, F. Analysis of population and economy of Maduo County in the source area of the Yellow River. *Popul. Res.* **1985**, *1*, 46–49. (In Chinese)
41. Piko, B.F.; Hamvai, C. Parent, school and peer-related correlates of adolescents' life satisfaction. *Child. Youth Serv. Rev.* **2010**, *32*, 1479–1482. [CrossRef]
42. Binder, M.; Coad, A. Life satisfaction and self-employment: A matching approach. *Small Bus. Econ.* **2013**, *40*, 1009–1033. [CrossRef]
43. Du, B.; Zhen, L.; Yan, H.; De Groot, R. Effects of Government Grassland Conservation Policy on Household Livelihoods and Dependence on Local Grasslands: Evidence from Inner Mongolia, China. *Sustainability* **2016**, *8*, 1314. [CrossRef]
44. Gao, L.; Kinnucan, H.W.; Zhang, Y.; Qiao, G. The effects of a subsidy for grassland protection on livestock numbers, grazing intensity, and herders' income in inner Mongolia. *Land Use Policy* **2016**, *54*, 302–312. [CrossRef]

45. Akay, A.; Martinsson, P.J.E.L. Does relative income matter for the very poor? Evidence from rural Ethiopia. *Econ. Lett.* **2008**, *110*, 213–215. [CrossRef]
46. Muresan, G.M.; Ciumas, C.; Achim, M.V. Can money buy happiness? Evidence for European countries. *Appl. Res. Qual. Life* **2020**, *15*, 953–970. [CrossRef]
47. *Leisure and Life Satisfaction: Foundational Perspectives;* Ginton, C.R. (Ed.) WCB/McGraw-Hill Press: New York, NY, USA, 1995.
48. Knight, J.; Lina, S.O.N.G.; Gunatilaka, R. Subjective well-being and its determinants in rural China. *China Econ. Rev.* **2009**, *20*, 635–649. [CrossRef]
49. Fujita, F.; Diener, E. Life satisfaction set point: Stability and change. *J. Personal. Soc. Psychol.* **2005**, *88*, 158–164. [CrossRef]

sustainability

Article

Citizenship Ability, Homestead Utility, and Rural Homestead Transfer of "Amphibious" Farmers

Zhongyou Yuan [1,2], Chenchen Fu [1], Shujie Kong [3], Jifeng Du [1,2] and Wei Li [1,2,4,*]

1 School of Public Administration, South China Agricultural University, Guangzhou 510642, China; yuanzhongyou@scau.edu.cn (Z.Y.); 15871741996@163.com (C.F.); dujifeng@scau.edu.cn (J.D.)
2 Key Laboratory of Natural Resources Monitoring in Tropical and Subtropical Area of South China, Guangzhou 510642, China
3 School of Public Affairs, Zhejiang University, Hangzhou 310058, China; 12022086@zju.edu.cn
4 Southern Marine Science and Engineering Guangdong Laboratory (Zhuhai), Sun Yat-sen University, Zhuhai 519000, China
* Correspondence: liwei289@sysu.edu.cn

Abstract: Rural homestead transfer is of considerable significance for the revitalization of rural land assets and sustainable use of land resources. "Amphibious" farmers are the most likely to transfer their homestead. As such, it is important to study their transfer behavior and influencing factors to promote homestead transfer. The study involved distributing questionnaires to 768 "amphibious" farmers in Guangdong Province, China, and 747 samples were valid. The impact of the farmers' citizenship ability and homestead utility preference on their homestead transfer behavior was analyzed using a binary logistic model. The study found that: (1) the citizenship ability of "amphibious" farmers had a significant positive impact on their willingness; the stronger the citizenship ability was, the stronger the willingness to transfer homesteads was. (2) Property and the guarantee utility of the homestead have an opposite impact on the willingness of "amphibious" farmers to transfer their homestead; if the property utility of the homestead is strong, its transfer intention is strong, but if the guarantee effect is strong, its transfer intention is weak. If the amphibious farmers are older, more educated, and have longer working years in cities, their willingness to transfer homestead will be lower. (3) The "amphibious" farmers working in cities and towns were found to have a stronger willingness to transfer homestead than those working in rural areas. Hence, the government should formulate differentiated policies for homestead transfer according to the ability endowment of farmers, improve various urban services and security infrastructure based on the urban housing of "amphibious" farmers, weaken the security utility of their rural homestead to flexibly realize their homestead property value, help them become citizens, and promote homestead transfer.

Keywords: "amphibious" farmers; willingness of homestead transfer; citizenship ability; homestead utility; binary logistic regression

Citation: Yuan, Z.; Fu, C.; Kong, S.; Du, J.; Li, W. Citizenship Ability, Homestead Utility, and Rural Homestead Transfer of "Amphibious" Farmers. *Sustainability* **2022**, *14*, 2067. https://doi.org/10.3390/su14042067

Academic Editor: Giuseppe Todde

Received: 6 December 2021
Accepted: 7 February 2022
Published: 11 February 2022

1. Introduction

Since the open reform, China's urbanization rate has increased from 17.92% in 1978 to 63.89% in 2020. Nearly 670 million farmers have moved to cities. This large-scale migration has led to significant changes in the relationship of urban and rural areas. Theoretically this will lead to the reduction of rural residential area and an increase in urban housing. However, the total size of homesteads has not decreased but increased with continuous population migration. Due to China's long-term implementation of the dual division of urban and rural areas and the urbanization of "diminishing land without transforming people" (hua di bu hua ren), a special type of farmer named "amphibious" farmers have appeared. On the one hand, amphibious farmers' registered residences are in rural areas, and they can own their rural homestead in the village. On the other hand, they have the ability to work and settle down in cities and purchase real estate. In earlier times, scholars defined amphibious farmers as "rural semi-urbanized floating people, working in the city and settled down in the countryside". They are engaged in agricultural activities in the farming seasons and engaged in non-agricultural activities such as industry, construction, transportation and commercial services in the slack seasons [1]. Later scholars found these farmers became social members with dual class attributes, who kept their identity as farmers and acted more like urban citizens [2]. Some defined amphibious farmers as "one identity and two occupations combing with agricultural and non-agricultural industries, working in the city with old houses in rural areas and a new home in the city" [3,4]. They are engaged less in agricultural activities and more in non-agricultural activities. Due to the Chinese household registration system, these farmers cannot easily change their registered residence in rural areas. This unique type of "amphibious" farmer has a dual land occupation in urban and rural areas, which has caused the imbalance between the supply and demand of urban construction land and housing, homesteads being left idle and inefficiently used, and illegal construction [5].

Existing research has realized the importance of homestead transfer in terms of farmers' attempts besides policy and government factors [6,7]. Zhang suggested that farmers should be guided and encouraged to withdraw their homestead from the perspective of farmers' benefits [8]. Zhao et al. investigated that the proportion of non-agricultural income in total income, the understanding of homestead policies and the education level have a positive impact on the farmers' attempts to transfer their homestead [9]. Xu and Liu found that the distance of homesteads from cities and towns has a negative impact on homestead transfer [10]. However, most of them consider farmers as a homogeneous population to investigate the influencing factors of farmers' attempts and pay less attention to classifying the farmers. Some research showed that farmers with non-agricultural employment have a different understanding of rural homestead transfer. Most of them work out of villages and do not consider their rural homestead as their only settlement, which provides more of a possibility to achieve the rural house transfer [11,12].

Therefore, it is meaningful to classify farmers to investigate the influencing factors of their attempts at homestead transfer. "Amphibious" farmers have a high chance of transferring their rural homestead. As the data is difficult to obtain, few studies have discussed it. In our research, we tried to focus on this special group of farmers and raised the following research question. What are the influencing factors of whether amphibious farmers transfer their rural homestead? This research tried to enrich the investigation of the differentiation of farmers' attempts at homestead transfer.

2. Literature Review

2.1. *Influencing Factors of Amphibious Farmers' Attempts at Transferring Their Rural Homestead*

Previous studies have discussed the influencing factors of farmers' attempts at transferring rural homestead mainly from two perspectives. The first perspective is the characteristics of farmers, such as the gender and age, status of non-agricultural employment, non-agricultural income and place of employment [13,14]. The second perspective is the function of the homestead. Lv and Zhao suggested that the function of the security and welfare of a homestead influence the issue of land compensation for farmers [15]. In the research, we followed these two perspectives and focused on amphibious farmers.

Firstly, the most significant characteristic of amphibious farmers is their citizenship ability, which is different from previous studies on homogeneous groups of farmers [4]. After 40 years of reform and opening the country, rural man–land relationships and urban–rural relationships in China have displayed new characteristics [16–19]. A survey shows that more and more rural migrant workers want to stay in cities, especially the new generation of migrant workers, and the proportion of them willing to settle in cities and towns is as high as 91.2% [20]. However, the citizenship intention is only the driving force behind citizenization. Whether it can be truly realized depends on citizenship ability, that is, the ability to settle and survive in cities and towns [21]. Research shows that only those farmers who were educated [13], had a significant proportion of non-agricultural income [22], and had a strong family livelihood [23], especially those who purchased houses in cities or towns, tended to give up rural homesteads and live in cities. In other words, only when urban life is guaranteed and the security needs of economic, social, and political aspects are met to a certain extent, will migrant workers consider giving up their land to completely become citizens [24], and realize the transformation from "farmers" to "migrant workers" and finally to "citizens" through citizenization. This shows that when farmers have a strong ability to settle and survive in cities, their demand for the guarantee function of homesteads will be weakened. Based on the above research, we supposed that there was a positive correlation between citizenship ability and attempts at homestead transfer by amphibious farmers.

Secondly, the function of homesteads is complex for amphibious farmers. The homestead is not only the basic material guarantee for farmers to live and work, and the social guarantee to prevent farmers from being displaced, but also the most important property basis for farmers to obtain more property income [25]. When the social guarantee system is weak, the rural homestead becomes the most important social guarantee for farmers [26]. However, with the non-agriculturalization of farmers' economic activities and the continuous improvement of their income level, as well as the increase in the proportion of farmers who buy houses in cities or towns, the importance of the residential function and living place of the homestead is decreasing, and the utility of farmers' free acquisition and occupation of the homestead is decreasing [27]. Furthermore, the focus of the homestead function has shifted, the guarantee function has been gradually weakened, and the property function has been gradually enhanced [28,29]. In the context of the change of homestead function, farmers will have different preferences for homestead function according to their own livelihood mode [30]. Those urban farmers with stable incomes and old-age security will hope to obtain income from land transfer and transfer their rural land [31]. Moreover, with the gradual weakening of the security attribute of the homestead and the continuous strengthening of the property attribute, the willingness of and appeal to farmers to transfer are becoming stronger [32]. Those farmers with urban settlement and living ability will abandon their rural land rights [33], while those farmers with a high dependence on their rural homesteads will be unwilling to quit the homestead. Based on the above analysis, we suppose that the property function of homesteads is positively correlated with homestead transfer and the guarantee function of homesteads is negatively related to the transfer of homesteads.

2.2. Theoretic Framework

We tried to find out the influencing factors of amphibious farmers' attempts at transferring their rural homestead from the above two perspectives and built up the theoretical framework shown in Figure 1.

One factor is citizenship ability, which consists of housing affordability, economic capability, and urban integration capability. The process of the citizenization of the rural population is not only reflected in career transformation [34], but also in citizen identity and urban integration [35]. Stable work and income in cities are the economic basis for amphibious farmers to integrate into the city [36], which can also promote their sense of identity as "urban citizens" [37]. Therefore, we consider higher economic income to play a positive role in promoting the urban social integration of amphibious farmers [38,39]. Moreover, housing is an intermediate mechanism for social division and integration [40]. It is not only a material space for social people to shelter, but also determines the living environment and social communication space of urban residents, providing opportunities for social people to obtain various urban resources, accumulate human and social capital, and integrate into the urban mainstream society [41,42]. Particularly for rural migrants, including amphibious farmers, owning urban housing will increase their willingness to stay in the city and promote their citizenization process [43,44].

The other influencing factor is homestead utility, including guarantee and property functions. The homestead is the place and guarantee for farmers to live and work, and it is the concrete carrier of farmers' concept of "home" [45]. Therefore, attempts at transferring rural homestead are very much affected by how owners think of their homestead's guarantee function. Moreover, the homestead is also the property of farmers. The wealth accumulation of a farmer's family from generation to generation is mainly reflected in the remaining "ancestral house" [46]. As the financial source of farmers, homesteads can not only obtain property income through leasing and equity investment, but also increase in value over time, bringing property income to farmers. If the homestead is expropriated, it can also obtain compensation income [47].

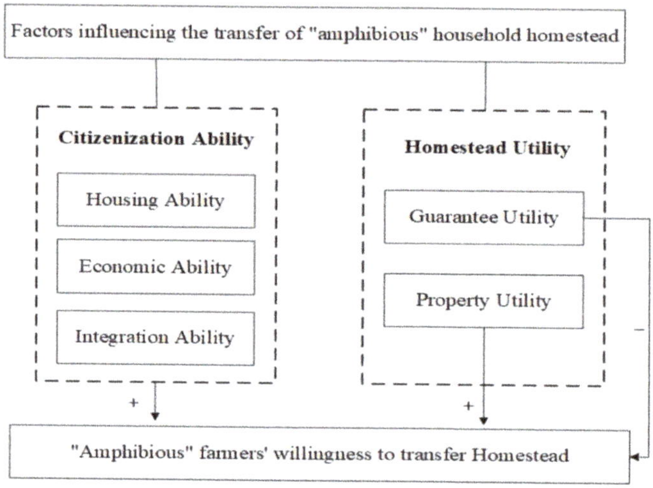

Figure 1. The theoretical analysis framework of the impact of citizenship and homestead utility on the willingness of "amphibious" farmers to transfer homestead. Note: "+" indicates improvement or enhancement, "−" indicates weakening or reduction.

However, with the further acceleration of urbanization and the citizenization of the agricultural transfer population, the homestead increasingly presents cashable potential functions, intergenerational inheritance functions, and land cultural and emotional functions [48]. Its core functions are still guarantee and property functions. Whether farmers transfer their homestead depends on the comparison of the potential benefits generated by guarantee functions and asset functions [49]. Especially since the No. 1 central document issued in 2018, the new policy of "three rights division" has been put forward. The balance between a homestead's function of guarantee and function of property is the key to whether the homestead is transferred. Concerning these two factors, we put forward three hypotheses.

Hypothesis 1 (H1). *The stronger the citizenship ability of amphibious farmers, the stronger their willingness to transfer homestead.*

Hypothesis 2 (H2). *The higher the property utility of amphibious farmers' homestead, the stronger their willingness to transfer homestead.*

Hypothesis 3 (H3). *The higher the guarantee utility of amphibious farmer households' homesteads, the lower their willingness to transfer their homestead.*

3. Material and Methods

3.1. Research Design

The research design includes four stages (Figure 2). The first stage is to suggest a hypothesis based on the literature review and research question. We planned to search the literature using key words of "amphibious" farmers and influencing factors to find out the research gap and theory foundation for the hypothesis.

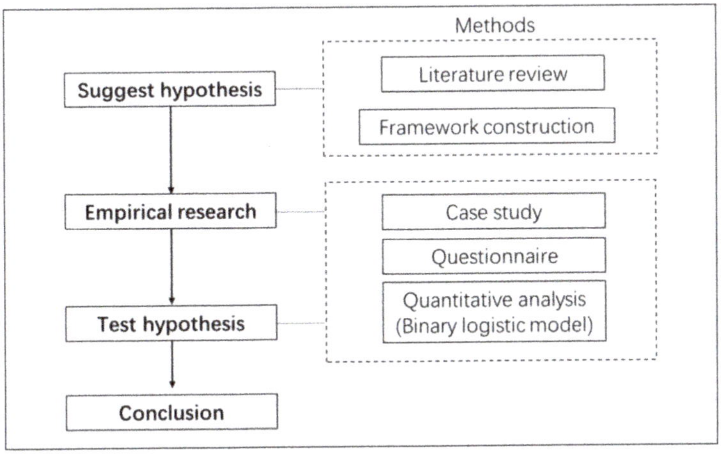

Figure 2. Research design and methods.

The second stage is empirical research. We planned to conduct a questionnaire (can be found in Appendices A and B) targeted at amphibious farmers' attempts at rural homestead transfer in a case area to obtain data. Amphibious farmers are a special group of migrant workers, and their distribution is random. It is difficult to identify them only when you know they have houses both in a city and in a village. They have their own social network which could be a special clue. Furthermore, it would be uneconomical and inefficient to adopt the usual sampling method as the group numbers and the proportion are not clear. Therefore, we planned to adopt snowball sampling to greatly increase the possibility of contacting this special group in the total population. The snowball sampling method starts with the survey objects familiar to the investigators, and then by obtaining more survey objects through them, like a snowball, gradually expand the scope of the sample until a reasonable sample size is obtained [50]. To prevent the inherent methodological limitation of snowball sampling to avoid biased results, we tried to do the sampling in different regions in the case area, which has a certain sample size and heterogeneity. Moreover, during the investigation, the triangular mutual verification method was adopted to ensure the quality of the data [51].

The third stage is to test the hypothesis with a quantitative analysis with data. The fourth stage comes to a conclusion and finds out the possible influencing factor of amphibious farmers' attempts at rural homestead transfer.

3.2. Case Study

We chose Guangdong province as a case study for two reasons. Firstly, amphibious farmers are more likely to appear in Guangdong province as its rural economy has developed fast and well. Guangdong province is at the forefront of opening up reform and is well developed in economic, social, and cultural aspects. Its industrialization and urbanization has developed well and lots of migrant workers from rural areas work in cities. According to the results of the third national agricultural census in Guangdong Province, there are 1,158,400 farmers purchasing real estate in cities or towns [52]. In 2021, the per capita disposable income of rural residents in Guangdong was CNY 22,306, an increase of 10.7%, much higher than that of the nation (CNY 18,931) [53]. Secondly, Guangdong province is also a pioneer area of reform of rural homestead transfer. Numerous regulations and policy practices have been implemented in Guangdong where farmers are open-minded and willing to answer questions on their private homestead and core profits.

3.3. Process of Survey in Questionnaire

We conducted the survey in Guangdong Province in 2019. The survey includes three steps. Step 1: from January to March 2019, a questionnaire was designed according to the research purpose of the subject. Step 2: April 2019, the pre-survey was carried out, and the final questionnaire was completed. Step 3: from May to August 2019, 32 experienced and trained investigators who are Guangdong registered residents and students returned to their hometowns to investigate "amphibious" farmers face-to-face. The respondents were mainly householders or their spouses, and each questionnaire was about 25 min long.

The questionnaire questions include farmers' personal and family characteristics, status of rural homestead, level of citizenship ability, and the willingness of the homestead transfer. The questionnaire adopted a 5-point Likert scale method, and the meaning of assignment ranges from "very poor" to "very good" (Appendices A and B). A total of 768 questionnaires were distributed. Due to it being a face-to-face survey, all the questionnaires were recovered, and the recovery rate was 100%.

3.4. Model Setting

The main purpose of this paper is to investigate "amphibious" farmers' homestead transfer intention and its influencing factors, taking the "amphibious" farmers' homestead transfer intention as the dependent variable Y, and the factors affecting the willingness of "amphibious" farmers to transfer their homestead as the independent variable X. The dependent variable "willingness of homestead transfer" has only two possibilities, namely "willing" and "unwilling" transfer, which is a binary variable. Therefore, the binary logistic regression model was used for analysis. By analyzing the relationship between Y and X, the influencing factors and contribution value of the willingness of "amphibious" farmers' homestead transfer can be clearly reflected. The basic form of the model is:

$$\ln\left(\frac{p(y=1)}{1-p(y=1)}\right) = \alpha + \sum_{i=1}^{k}(\beta_i x_i) \tag{1}$$

Then the probability of homestead transfer by "amphibious" farmers is:

$$p(y=1|x) = \frac{e^{\alpha+\sum_{i=1}^{k}\beta_i x_i}}{1+e^{\alpha+\sum_{i=1}^{k}\beta_i x_i}} \tag{2}$$

In Formulas (1) and (2), X_i represents the i explanatory variable affecting the willingness for homestead transfer by "amphibious" farmers and k is the number of explanatory variables, α is the intercept term, and β_i is the coefficient of the explanatory variable X_i, reflecting the direction and degree of the impact of the variable on the willingness for homestead transfer by "amphibious" farmers, which is usually obtained by the maximum likelihood estimation method. The ratio $\frac{p(y=1)}{1-p(y=1)}$ is the event occurrence ratio, which is the probability of "amphibious" farmers being willing to transfer homestead or unwilling to transfer homestead, and e^{β_i} reflects the multiple of the change of the event occurrence ratio caused by each unit of the explanatory variable X_i.

In order to make the results of empirical analysis more robust and reliable in the binary logistic analysis, the core explanatory variables were included in three steps to construct the analysis model. After the control variables were included, only the core explanatory variables such as guarantee utility were included in Model (1). Model (2) included two core explanatory variables: guarantee utility and property utility after the control variables were included. Model (3) included three core explanatory variables: guarantee utility, property utility, and citizenship ability after the control variables were included.

4. Results

4.1. Descriptive Statistics by "Amphibious" Farmers' Willingness of Homestead Transfer

The valid survey data numbered 747, and the characteristics were as follows: (1) Gender distribution: there were 422 men, accounting for 56.49% of the total; there were 325 women, accounting for 43.51% of the total. (2) Age distribution: there were 115 people aged 20–30, accounting for 15.39% of the total; 204 people aged 31 to 40, accounting for 27.31% of the total; 291 people aged 41 to 50, accounting for 38.96% of the total; 107 people aged 51 to 60, accounting for 14.32% of the total; and 30 people over the age of 61, accounting for 4.02% of the total. (3) Education distribution: there were 81 people with primary school education and below, accounting for 10.84% of the total; 216 people with junior middle school education, accounting for 28.92% of the total; 163 people with high school education, accounting for 21.82% of the total; 134 people with specialized education, accounting for 17.94% of the total; and 153 people with bachelor's degrees or above, accounting for 20.48% of the total. (4) Annual household income distribution: 20 households earned below CNY 20,000, accounting for 2.68% of the total; 99 households with CNY 20,000 to 60,000, accounting for 13.25% of the total; 184 households with CNY 70,000 to 110,000, accounting for 24.63% of the total; 229 households with CNY 120,000 to 160,000, accounting for 30.66% of the total; and 215 households above 160,000 yuan, accounting for 28.78% of the total. (5) Regional distribution: there were 215 households in northern Guangdong, accounting for 28.78% of the total; 243 households in the east and west wings, accounting for 32.53% of the total; and 289 households in the Pearl River Delta, accounting for 38.69% of the total.

Table 1 presents the reliability test results of measurement indicators. The calculation results show that the Cronbach's alpha value of homestead utility is greater than 0.6, and the Cronbach's alpha value of citizenship ability is greater than 0.7, indicating that the variable has good reliability. In addition, the principal component analysis was carried out on 13 observed variables in guarantee utility, property utility, economic ability, integration ability, and housing ability; the KMO value of the model is 0.700; and the Bartlett's spherical test values were significant at the statistical standard of 1%.

80 per cent of "amphibious" farmers were satisfied with the living environment and conditions of their rural houses. A total of 74.83% of "amphibious" farmers think homesteads can give them a sense of family belonging. Secondly, in terms of property utility in homestead utility, 62.11% of "amphibious" farmers agree that compensation for homestead demolition and requisition will bring them greater economic benefits, and 10.32% of "amphibious" farmers thought that homestead demolition and requisition will not bring them greater economic benefits. More than half (54.48%) of the "amphibious" farmers believe that their homestead can obtain greater benefits in the process of transfer, and about 17.01% of the "amphibious" farmers believe that homestead transfer will not bring them greater benefits. Thirdly, in terms of economic ability in citizenship ability, 59.43% of the "amphibious" farmers have an annual income of more than CNY 110,000, indicating that the annual income of the "amphibious" farmers is in a good condition. However, only 27.00% of the "amphibious" farmers accepted the price levels in the city where they live. In total, 85.00% of "amphibious" farmers believe that their non-agricultural livelihood skills are sufficient or able to cope with urban life. Fourthly, in terms of integration ability in citizenship ability, nearly half of "amphibious" farmers (nearly 46.31%) have some difficulties in integrating into the new urban life and still believe that (68.13%) they have the strong ability to get involved in urban life. Fifthly, in terms of housing capacity in citizenship ability, up to 95.00% of "amphibious" farmers are satisfied or relatively satisfied with the living environment, public supporting facilities, and community facilities of their urban houses, and only 5.00% of "amphibious" farmers are dissatisfied with the living environment of their urban houses.

Table 1. Reliability test results of measurement indicators.

Criterion Layer	Latent Variable	Variable	Mean	SD	Item's Deleted Alpha Value	Cronbach's Alpha Value
Homestead utility	Guarantee utility	Satisfaction with rural house living environment	3.31	0.79	0.609	0.638
		Satisfaction with rural house living conditions	3.22	0.89	0.617	
		The homestead with a sense of family belonging	3.99	0.88	0.602	
	Property utility	Large compensation value for demolition and requisition	3.78	0.99	0.532	
		Large income from homestead transfer	3.61	1.12	0.549	
Citizenship ability	Economic ability	Annual household income (CNY 10,000)	3.70	1.10	0.685	0.707
		Acceptance of urban prices	2.85	0.91	0.702	
	Integration ability	Non-agricultural livelihood skills	3.18	0.75	0.684	
		Compared with the self-feeling of urban residents	2.92	0.98	0.689	
		Acceptance of new urban life	3.51	0.80	0.696	
	Housing ability	Satisfaction with urban housing	3.59	0.77	0.660	
		Satisfaction with public supporting facilities of urban housing	3.49	0.82	0.649	
		Satisfaction with urban housing community facilities	3.50	0.83	0.655	

4.2. Binary Logistic Results on "Amphibious" Farmers' Willingness of Homestead Transfer

The test results show that the maximum variance expansion factor (VIF) of the selected explanatory variables is 2.255 and the minimum is 1.134, which can determine that there is no collinearity problem between the explanatory variables of the model.

Table 2 presents estimation results of the binary logistic model. According to the fitting information of the regression model, the omnibus test results of the model coefficients show that the chi square values of Model (1), (2), and (3) are 32.964, 56.496, and 123.683, respectively, and the P values are less than 0.000. The Hosmer–Lemeshow goodness of fit test results show that the chi square values of Model (1), (2), and (3) are 17.186, 8.947, and 7.171, respectively, and the P values are 0.124, 0.347, and 0.518, respectively, which are greater than 0.05, indicating that the model has a goodness of fit and is suitable for the binary logistic regression model. The direction and size of independent variables can be analyzed and judged by regression results.

Among the 747 "amphibious" farmers in Guangdong Province, 225 of them actually transferred their homesteads, accounting for 30.12%; this rate is higher than the homestead transfer rate of ordinary farmers in Wuhan (18.97%), surveyed by Guan and Huang (2013), and the homestead transfer rate of ordinary farmers in Wenzhou (13.15%), surveyed by Qian et al. (2015), indicating that "amphibious" farmers' willingness to transfer their homestead is higher than that of ordinary farmers.

The estimation results of Model (3) including all variables show that under the control of individual characteristic variables of farmers, six of the eight variables that measure the citizenship ability of "amphibious" farmers have a positive impact on their willingness to transfer their homestead; among them, non-agricultural livelihood skills (β = 0.782, OR = 2.186, $p < 0.01$), satisfaction with urban housing (β = 0.312, OR = 1.367, $p < 0.05$), satisfaction with public supporting facilities of urban housing (β = 0.567, OR = 0.567, $p < 0.01$), and the satisfaction with urban housing community facilities (β = 0.570, OR = 1.768, $p < 0.01$) had a significant positive impact on their willingness to transfer their homestead. For each unit of "amphibious" farmers' non-agricultural livelihood skills, satisfaction with urban housing, satisfaction with public supporting facilities of urban housing, and satisfaction with urban housing community facilities, their willingness to transfer homestead increased by 2.186, 1.367, 0.567, and 1.768 times, respectively. It shows that the stronger the citizenship ability of "amphibious" farmers, the stronger their willingness to transfer homestead. Hypothesis 1 has been verified.

Table 2. Estimation results of binary logistic model.

Variable Category			Model (1)		Model (2)		Model (3)	
			β	**Exp(β)**	**β**	**Exp(β)**	**β**	**Exp(β)**
Homestead utility	Guarantee utility	Satisfaction with rural house living environment	−0.330 ***	1.391	−0.311 **	1.365	−0.237 *	1.267
		Satisfaction with rural house living conditions	−0.158	0.854	−0.171	0.843	−0.215 *	0.806
		The homestead with a sense of family belonging	−0.095	0.909	−0.235 **	0.791	−0.198 *	0.821
	Property utility	Large compensation value for demolition and requisition			−0.108	1.114	0.073	1.075
		Large income from homestead transfer			0.323 ***	1.381	0.375 ***	1.455
Citizenship ability	Economic ability	Annual household income (CNY 10,000)					−0.319 ***	0.727
		Acceptance of urban prices					0.156	1.169
	Integration ability	Non−agricultural livelihood skills					0.782 ***	2.186
		Compared with the self feeling of urban residents					0.025	1.025
		Acceptance of new urban life					−0.037	0.963
	Housing ability	Satisfaction with urban housing					0.312 **	1.367
		Satisfaction with public supporting facilities of urban housing					0.567 ***	0.567
		Satisfaction with urban housing community facilities					0.570 ***	1.768
Control Variable	Gender (male as control group)		−0.214	0.808	−0.229	0.795	−0.269	0.764
	Age		−0.198 **	0.821	−0.205 **	0.814	−0.204 **	0.815
	Work location (compared with cities and towns)		−0.287 **	0.751	−0.229 *	0.795	−0.249 *	0.780
	Education		−0.184 **	0.832	−0.214 ***	0.807	−0.235 ***	0.790
	Years of working in cities		−0.114 *	0.892	−0.123 *	0.884	−0.136 *	0.873
Constant			1.192	3.293	0.277	1.319	−2.315	0.099

Note: (1) *, **, *** are significant at the level of 10%, 5%, and 1%, respectively; (2) β is the regression coefficient, "−" is the negative influence; (3) EXP(β) is the OR value, also known as odds ratio or occurrence ratio.

At the same time, Model (3) estimation results show that the two core explanatory variables to measure the property utility of "amphibious" farmers' homesteads have a positive effect on the willingness of "amphibious" farmers to transfer their homestead, among which the income of homestead transfer (β = 0.375, OR = 1.455, $p < 0.01$) has a significant positive impact on their willingness to transfer their homestead. For each unit, the willingness of "amphibious" farmers to transfer increases by 1.455 times. This shows that the stronger the property effect of a homestead, the more willing "amphibious" farmers are to transfer their homestead, and Hypothesis 2 is verified.

Model (3) estimation results also show that the satisfaction with rural house living environment (β = −0. 237, OR = 1.267, $p < 0.1$), satisfaction with rural house living conditions (β = −0.215, OR = 0.806, $p < 0.1$), and a sense of family belonging (β = −0. 198, OR = 0.821, $p < 0.1$)—the three core explanatory variables that measure the guarantee utility of a homestead—have a significant negative impact on the willingness of "amphibious" farmers to transfer their homestead. For each unit of "amphibious" farmers' satisfaction with the rural house living environment, the rural house living conditions, and the homestead with a sense of family belonging, their willingness to transfer homestead decreased by 1.267, 0.806, and 0.821 times, respectively. This shows that the stronger the guarantee effect of a homestead, the more reluctant "amphibious" farmers are to transfer their homestead, and Hypothesis 3 has been verified.

From the individual characteristics of "amphibious" farmers, the age of "amphibious" farmers (β = −0.204, OR = 0.815, $p < 0.05$), education of head of household (β = −0. 235, OR = 0.790, $p < 0.01$), and years of working in cities (β = −0.136, OR = 0.873, $p < 0.1$), and so on, have a significant negative impact on their homestead transfer. The older the "amphibious" farmers are, the more educated they are, and the longer they have been working in cities, the more reluctant they are to transfer their homestead; for each unit with an increase in age, education level, and years of working in cities, their willingness to transfer decreases by 0.815, 0.790, and 0.873 times, respectively. A possible explanation is that the overall income of rural homestead transfer is low. For "amphibious" farmers with a high annual family income, the income brought in by homestead transfer accounts for a small proportion of their family income, and they pay more attention to the intergenerational inheritance function of homesteads and the emotional function of land culture [10]. Highly educated "amphibious" farmers have a better understanding of the current rural policies and the scarcity of land, think that the homestead has appreciation potential, are unwilling to transfer their homestead cheaply, and thus have a wait-and-see attitude towards the transfer of their homestead. The older "amphibious" farmers have deeper feelings for and stronger dependence on rural areas and homesteads, so their willingness to transfer is lower than that of young farmers. Working in the countryside (β = −0. 249, OR = 0. 780, $p < 0.1$) has a significant negative impact on the homestead transfer of "amphibious" farmers. The "amphibious" farmers working in cities or towns are more willing to transfer their homestead than the "amphibious" farmers working in rural areas. The possible explanation is that the "amphibious" farmers who work in cities or towns all year round have adapted to urban life and are unwilling to return to the countryside, hoping to transfer their homestead and realize the property value of their homestead, thus helping in becoming an urban resident.

5. Discussion

Meanings of Citizenship Ability and Homestead Efficiency Preference of "Amphibious" Farmers

Most research considers farmers to be a homogeneous population to discuss its influencing factors of attempts at rural homestead transfer [8–10]. In our survey, we argued that with the differentiation of occupation and income, it is necessary to consider "amphibious" farmers as a special group based on its dual homestead occupation in rural and urban areas [54,55]. It was found that the potential rate of homestead transfer in the groups of "amphibious" farmers in Guangdong Province is 30.12%. This is much higher than other surveys of non-amphibious farmers in Wuhan (13.15%) [56] and Wenzhou cases (13.15%) [57,58]. During the process of rural industrialization and urbanization, the rural–urban linkages become more and more diverse and complex. Amphibious farmers are representative actors to make rural–urban linkages as they have properties both in urban and rural areas. Therefore, when we discuss rural homestead transfer, it is important to classify the groups of amphibious farmers and consider their interests not only in rural villages, but also in towns and cities.

The research attempted to find out the influencing factors of amphibious farmers' attempts at rural homestead transfer. According to our results, we found a positive correlation between citizenship ability and attempts at homestead transfer by amphibious farmers. In addition, the property function of homesteads is positively correlated with homestead transfer, and the guarantee function of homesteads is negatively related to the transfer of homesteads.

Some factors seem similar to non-amphibious farmers, such as the positive effect of farmers' citizenship abilities. Peng and Liu found that farmers who have a stronger ability to settle down in a city have more willingness to transfer their rural homestead [18,22]. In addition, we have proven the hypothesis that amphibious farmers are inclined to weaken the security function of their rural homestead because they have the ability to settle down in cities and no need to maintain their rural homestead.

However, some findings contrast ours. Xu and Liu found that farmers who are educated are more likely to transfer their rural homestead as they are open-minded and understand the meaning of the policy of rural homestead transfer [10]. In contrast, we found that the more educated amphibious farmers are, the less willing they are to transfer rural homestead. Actually, a rural homestead has multiple functions, such as security, culture, and ecology [59]. Educated amphibious farmers may evaluate rural homesteads not only from the physical side, but also from the cultural and spiritual side [58]. Thus, they evaluate their rural homestead for more potential profits, and are not willing to transfer with less money. In other words, when negotiating with amphibious farmers, it would cost much more compensation money than non-amphibious farmers, which previous studies easily neglected.

6. Conclusions

Rural homestead transfer is beneficial for increasing the efficiency of rural land use and rural sustainability. Though the existing research realized the importance of farmers' role in homestead transfer, they paid less attention to classifying groups of farmers. This study tried to discuss farmers' attempts at rural homestead transfer from a typical and specific group of "amphibious" farmers, who own homesteads both in rural and urban areas and have a higher potential to transfer their rural homestead, which highlights the importance of the classification of farmers when analyzing their attempts at rural homestead transfer.

Furthermore, the study revealed that citizenship ability and preference of homestead utility affect their transfer behavior through binary logistic analysis on 747 valid cases in Guangdong province. This study proved the three hypotheses that the stronger the citizenship ability is, the stronger the willingness to transfer homesteads is; if the property utility of a homestead is stronger, its transfer possibility is stronger; and if the guarantee utility of a homestead is stronger, its transfer possibility is stronger. This study also found that the willingness of "amphibious" farmers to transfer their homestead is also affected by individual and family characteristics. The older the "amphibious" farmers are, the more educated they are, and the longer they have been working in cities, the more reluctant they are to transfer their homestead. Amphibious farmers working in cities or towns are more willing to transfer their homestead than those working in rural areas.

We suggest the current policy should be guided to affect citizenship ability and preference of homestead utility to influence the willingness of amphibious farmers' rural homestead transfer. In the future, we will make more comparable studies to distinguish drivers and factors among farmers with different characteristics.

Author Contributions: Conceptualization, Z.Y. and W.L.; methodology, Z.Y.; validation, Z.Y., C.F. and S.K.; formal analysis, S.K. and J.D.; investigation, Z.Y. and J.D.; resources, Z.Y. and W.L.; writing—original draft preparation, Z.Y. and S.K.; writing—review and editing, Z.Y. and W.L.; project administration, Z.Y. and W.L.; funding acquisition, Z.Y. and W.L. All authors have read and agreed to the published version of the manuscript.

Funding: This research was funded by the Ministry of Education Humanities and Social Sciences Research Planning Fund, grant number 18YJAZH121; the general project of "Third-tenth Five-Year Plan" in 2018 for Guangdong Province Philosophy and Social Sciences, grant number GD18CGL11. The research was also funded by the National Natural Science Foundation of China, 41901192, and supported by the Key Laboratory of Natural Resources Monitoring in Tropical and Subtropical area of South China.

Institutional Review Board Statement: Not applicable.

Informed Consent Statement: Informed consent was obtained from all subjects involved in the study.

Conflicts of Interest: The authors declare no conflict of interest.

Appendix A. Core Independent Variables and Their Assignment in Questionnaire

Variable Name			Variable Assignment
Citizenship Ability	Economic Ability	Annual household income (CNY 10,000)	1 = 2 or less, 2 = 2–6, 3 = 7–11, 4 = 12–16, 5 = 16 and above
		Acceptance of urban prices	1 = Totally unacceptable, 2 = Less acceptable, 3 = Commonly, 4 = Quite acceptable, 5 = Totally acceptable
	Integrate Ability	Non-agricultural livelihood skills	1 = Very weak, 2 = Relatively weak, 3 = Commonly, 4 = Relatively skilled, 5 = very skilled
		Compared with the self-feeling of urban residents	1 = Much worse than them, 2 = Worse than them, 3 = Equally, 4 = Better than them, 5 = Much better than them
		Acceptance of new urban life	1 = Very weak, 2 = Relatively weak, 3 = Commonly, 4 = Relatively skilled, 5 = Very skilled
	Housing Ability	Satisfaction with urban housing	1 = Very dissatisfied, 2 = Quite dissatisfied, 3 = Commonly, 4 = Quite satisfied, 5 = Very satisfied
		Satisfaction with public supporting facilities of urban housing	1 = Very dissatisfied, 2 = Quite dissatisfied, 3 = Commonly, 4 = Quite satisfied, 5 = Very satisfied
		Satisfaction with urban housing community facilities	1 = Very dissatisfied, 2 = Quite dissatisfied, 3 = Commonly, 4 = Quite satisfied, 5 = Very satisfied
Homestead Utility	Guarantee Utility	Rural house living environment	1 = Very dissatisfied, 2 = Quite dissatisfied, 3 = Commonly, 4 = Quite satisfied, 5 = Very satisfied
		Rural house living conditions	1 = Very dissatisfied, 2 = Quite dissatisfied, 3 = Commonly, 4 = Quite satisfied, 5 = Very satisfied
		The homestead with a sense of family belonging	1 = Totally disagree, 2 = Comparative disagree, 3 = commonly, 4 = More agree, 5 = Fully agree
	Property Utility	Large compensation value for demolition and requisition	1 = Totally disagree, 2 = Comparative disagree, 3 = Commonly, 4 = More agree, 5 = Fully agree
		Large income from homestead transfer	1 = Totally disagree, 2 = Comparative disagree, 3 = Commonly, 4 = More agree, 5 = Fully agree

Appendix B. Core Independent Variables and Their Assignment in Questionnaire

Variable Name			Variable Assignment
Citizenship Ability	Economic Ability	Annual household income (CNY 10,000)	1 = 2 or less, 2 = 2–6, 3 = 7–11, 4 = 12–16, 5 = 16 and above
		Acceptance of urban prices	1 = Totally unacceptable, 2 = Less acceptable, 3 = Commonly, 4 = Quite acceptable, 5 = Totally acceptable
	Integrate Ability	Non-agricultural livelihood skills	1 = Very weak, 2 = Relatively weak, 3 = Commonly, 4 = Relatively skilled, 5 = very skilled
		Compared with the self-feeling of urban residents	1 = Much worse than them, 2 = Worse than them, 3 = Equally, 4 = Better than them, 5 = Much better than them
		Acceptance of new urban life	1 = Very weak, 2 = Relatively weak, 3 = Commonly, 4 = Relatively skilled, 5 = Very skilled
	Housing Ability	Satisfaction with urban housing	1 = Very dissatisfied, 2 = Quite dissatisfied, 3 = Commonly, 4 = Quite satisfied, 5 = Very satisfied
		Satisfaction with public supporting facilities of urban housing	1 = Very dissatisfied, 2 = Quite dissatisfied, 3 = Commonly, 4 = Quite satisfied, 5 = Very satisfied
		Satisfaction with urban housing community facilities	1 = Very dissatisfied, 2 = Quite dissatisfied, 3 = Commonly, 4 = Quite satisfied, 5 = Very satisfied
Homestead Utility	Guarantee Utility	Rural house living environment	1 = Very dissatisfied, 2 = Quite dissatisfied, 3 = Commonly, 4 = Quite satisfied, 5 = Very satisfied
		Rural house living conditions	1 = Very dissatisfied, 2 = Quite dissatisfied,3 = Commonly, 4 = Quite satisfied, 5 = Very satisfied
		The homestead with a sense of family belonging	1 = Totally disagree, 2 = Comparative disagree, 3 = commonly, 4 = More agree, 5 = Fully agree
	Property Utility	Large compensation value for demolition and requisition	1 = Totally disagree, 2 = Comparative disagree, 3 = Commonly, 4 = More agree, 5 = Fully agree
		Large income from homestead transfer	1 = Totally disagree, 2 = Comparative disagree, 3 = Commonly, 4 = More agree, 5 = Fully agree

References

1. Yao, S.; Wu, C. A special form of urbanization of rural population in China—On China's industrial and agricultural population. *J. Geogr.* **1982**, *37*, 155–163.
2. Yang, Y. Shortage of migrant workers, semi urbanization model and urbanization model. *Economist* **2010**, *9*, 71–76.
3. Fu, G. From amphibian to triphibian: Change of migrant workers' life way and its influence. *J. Northwest AF Univ.* **2018**, *18*, 31–36, 44. [CrossRef]
4. Wang, Y.; Yan, Z.; Zhu, W.; Wang, H. Study on influencing factors and mode of action of amphibious farmers in parting with their lands in Henan. *J. Shandong Acad. Agric. Eng.* **2016**, *33*, 1–7. [CrossRef]
5. Zhang, Y.; Zhou, L.; Jia, W. Research progress and prospects on revitalization andutilization of rural residential land. *J. China Agric. Univ.* **2020**, *25*, 129–141. [CrossRef]
6. Zhang, Y.; Westlund, H.; Klaesson, J. Report from a Chinese Village 2019: Rural Homestead Transfer and Rural Vitalization. *Sustainability* **2020**, *12*, 8635. [CrossRef]
7. Long, H. Land Use Transition and Rural Transformation Development. *Prog. Geogr.* **2012**, *31*, 131–138. [CrossRef]
8. Zhang, Y. Policy Combing and Practice Comparison of Rural Homestead Withdraw with Compensation Since 2015. *J. Northwest AF Univ.* **2019**, *19*, 83–89. [CrossRef]
9. Zhao, G.; Yang, G. Analysis on Farmer's Attention of the Rural Homestead Conversion and its Influence Factors—Based on the Farm Household in Two Counties of Hubei Province. *Resour. Environ. Yangtze Basin* **2009**, *18*, 1121–1124.
10. Xu, H.; Liu, C. Research on Farmers Use House stead Circulation Wishes and Affecting Factors Based on Wuhan Jiangxis District 210 Households Questionnaire Analysts. *J. Northwest AF Univ.* **2012**, *12*, 44–49. [CrossRef]
11. Qian, L.; Qian, W.; Chen, F. Farmers' Differentiation, Expectations of Property and Rural Housing Land Transference: Based on the Survey and Empirical Analysis of Wenzhou. *China Land Sci.* **2015**, *29*, 19–26. [CrossRef]
12. Peng, C.; Wang, Q.; Zhong, Y. Rural Land Right Confirmation, Farmers' Differentiation and Willingness of Homestead Disposal: An Empirical Analysis Based on Survey Data of Anhui and Hunan Provinces. *J. Nanjing Agric. Univ.* **2019**, *19*, 118–129, 158. [CrossRef]
13. Xie, Y.; Jiang, Q. Land arrangements for rural–urban migrant workers in China: Findings from Jiangsu Province. *Land Use Policy* **2016**, *50*, 262–267. [CrossRef]
14. Lv, J.; Zhang, S. The formation factors of amphibious land occupation in rural and urban areas, the obstacles of land withdrawal system and policy suggestions—Sample analysis based on hundred village survey. *Reform Econ. Syst.* **2020**, *2*, 66–73.

15. Zhao, B. Institutional obstacles and solutions to the "amphibious land occupation" of farmers in urban and rural areas. *Agric. Econ.* **2020**, *3*, 68–70.
16. Guan, J.; Huang, C. Microscopic Welfare and Risk Perspective of Rural Residential Land Circulation: Wuhan Survey. *Reform* **2013**, *8*, 78–85.
17. Fishbein, M.; Ajzen, I. Belief, Attitude, Intention and Behavior: An Introduction to Theory and Research. *Philos. Rhetor.* **1977**, *10*, 130–132. [CrossRef]
18. Breckler, S.J.; Wiggins, E.C. Affect versus Evaluation in the Structure of Attitudes. *J. Exp. Soc. Psychol.* **1989**, *25*, 253–271. [CrossRef]
19. Zhou, J.; Yang, Q. Review of the Researches on Rural Housing Land Transfer at Farm Household Leve. *Prog. Geogr.* **2012**, *31*, 139–148. [CrossRef]
20. Liu, S. Land Issues in Urban-Rural China. *J. Peking Univ. Philos. Soc. Sci.* **2018**, *55*, 79–93.
21. Research Group of Development Research Centre of the State Couneil. The General Situation and Strategy Orientation of the Citizenization Process of the Migrant Workers. *Reform* **2011**, *5*, 5–29.
22. Liu, T.; Zhang, Y.; Kong, X. The Citizenship ability, Land Rights Cognition and Farmers' Willingness to Quitting Land. *China Land Sci.* **2013**, *27*, 23–30. [CrossRef]
23. Yang, J. Empirical analysis of factors influencing farmers' willingness to transfer homestead—Take Xitang town of Jieyang city in Guangdong province as an example. *J. Zhongkai Univ. Agric. Eng.* **2016**, *29*, 54–58. [CrossRef]
24. Chen, M.; Kuang, F.; Lu, Y. Livelihood Capital Differentiation and Farmers' Willingness to Homestead Circulation: Based on Empirical Analysis of Jiangxi Province. *J. Agro For. Econ. Manag.* **2018**, *17*, 82–90. [CrossRef]
25. Fan, J.; Mo, J. Urbanization Style and the Pattern of Economic Development: Also on the Direction of Urbanization. *Fudan J. Soc. Sci.* **2013**, *55*, 65–73, 167.
26. Xu, Z.; Zhuo, Y.; Wu, C.; Li, G. Review on Rural Homestead Studies. *Issues Agric. Econ.* **2019**, *4*, 28–39. [CrossRef]
27. Chen, B. Rationality of Limitations on Free Transaction of Rural Residential Land. *China Land Sci.* **2007**, *4*, 44–48. [CrossRef]
28. Liu, S.; Xiong, X. Changes in Economic Structure, Village Transformation and Changes in Homestead System: A Case Study of Homestead System Reform in Luxian County, Sichuan Province. *Chin. Rural Econ.* **2018**, *6*, 2–20.
29. Zhang, K.; Fu, Z. Based on the Functional Change of the Exploration of the Land System Reform. *Soc. Sci. Res.* **2017**, *6*, 47–53.
30. Li, R.; Ye, X. Exit and Flow: Farmers' Land Disposal Choice and Influencing Factors. *Rural Econ.* **2019**, *4*, 10–20.
31. Wang, J. Rural Homestead Function and Price of Empirical Research in China, Based on the Analysis of the Function of Different Types of Peasant Household Land Preference. *Price Theory Pract.* **2016**, *7*, 93–96. [CrossRef]
32. He, X.; Huang, X. Farmland circulation can never "irreversible". *Land Resour. Her.* **2011**, *8*, 52–55.
33. Dong, X. The Circulation of Housing Land Use Right in the Past 70 Years Since the Founding of New China: Institutional Change, Current Dilemma and Reform Direction. *Chin. Rural Econ.* **2019**, *6*, 2–27.
34. Chen, H.; Liu, Z.; Shi, X. The Role of Land Rights and Interests on Migrant Workers' Urban-rural Migration Decision: A Case Study on 1062 Migrant Workers in a Sampling Survey of Nanjing City. *Issues Agric. Econ.* **2012**, *33*, 70–77, 111–112. [CrossRef]
35. Borjas, G.J. Self-Selection and the Earnings of Immigrants. *Am. Econ. Rev.* **1987**, *77*, 531–553.
36. Liu, C.; Zhou, L. The Urban Integration of Social Capital and Migrant Workers. *Popul. Res.* **2004**, *5*, 12–18.
37. Liu, Y.; Yang, Q. Research Advance on Farmers' Land Property Income under the Background of New Urbanization in China. *Econ. Geogr.* **2019**, *39*, 164–171. [CrossRef]
38. Chen, J.; Yang, X. The Study of the Index System for Ability Evaluation on Fusing into Cities and Towns by Migrant Workers. *Urban Probl.* **2012**, *8*, 58–63. [CrossRef]
39. Zhang, X. The Research of improving the new generation of urban migrant workers into the ability. *Guizhou Soc. Sci.* **2011**, *7*, 79–82. [CrossRef]
40. Shi, Z.; Zhu, M. Employment Stability and Social Integration of Migrant Workers. *J. Zhong Nan Univ. Econ. Law* **2014**, *3*, 49–58, 159. [CrossRef]
41. Algan, Y.; Dustmann, C.; Glitz, A.; Manning, A. The Economic Situation of First and Second-Generation Immigrants in France, Germany and the United Kingdom. *Econ. J.* **2010**, *120*, 4–30. [CrossRef]
42. Hamermesh, D.S.; Trejo, S.J. How do immigrants spend their time? The process of assimilation. *J. Popul. Econ.* **2013**, *26*, 507–530. [CrossRef]
43. Zheng, S.; Liao, J.; Ren, R.; Cao, Y. Housing Policy for Migrant Workers and Economic Growth. *Econ. Res. J.* **2011**, *46*, 73–86.
44. World Bank. *World Development Report 1995*; China Financial &Economic Publishing House: Beijing, China, 1995.
45. Zhu, Z.; Leng, C. Housing Conditions, Social Status and Urban Identity of Migrant Workers: An Empirical Analysis Based on Social Integration Survey Data. *China Rural. Surv.* **2018**, *1*, 96–110.
46. Li, Y.; Liu, N.; Li, X. Farmland Circulation, Housing Choice and Peasant-Workers' Citizenization. *Econ. Geogr.* **2019**, *39*, 165–174. [CrossRef]
47. Yang, Q.; Chen, C. Economic Agglomeration, Housing Affordability and Migration Willingness of the Floating Population. *Mod. Financ. Econ. J. Tianjin Univ. Financ. Econ.* **2019**, *39*, 29–45. [CrossRef]
48. The Research of the Rural Land System Based on the Rural Hollowing out the Background Team. The Function Reconfiguration of House-site in the Countryside in the Process of Urbanization. *Rural Econ.* **2016**, *4*, 15–19.
49. Chen, X.; Jiang, X. The System of the Right to Use the Land for the House Foundation: An Analysis of its Norm, the Challenge in Practice and the Legislative Response. *Manag. World* **2010**, *12*, 1–12. [CrossRef]

50. Chen, L.; Wang, Q.; Long, K. The Theoretical Analysis of the Influential Factors of Welfare Level by Farmers' land. *Rural Econ.* **2011**, *12*, 13–16.
51. Wang, J.; Xue, Y. How to grasp the meaning in social science—Research on interpretation method based on triangulation. *Theorists* **2012**, *9*, 85–87. [CrossRef]
52. Main Data Bulletin of the Third National Agricultural Census of Guangdong Province. Available online: http://stats.gd.gov.cn/tjgb/content/post_1430129.html (accessed on 5 February 2021).
53. Guangdong Provincial Bureau of Statistics. *Guangdong Statistical Yearbook 2021*; China Statistics Press: Beijing, China, 2021.
54. Zhang, Y.; Xu, C. The compensation value composition of exiting rural residential land in the multifunctional perspective. *J. Shanxi Agric. Univ.* **2018**, *17*, 22–27. [CrossRef]
55. Peng, C. Analysis of the Influence of Peasants' Social Stratum on their Choice of Monetary Compensation Model for Leaving Their Rural Homestead: Based on Survey Data of Farmers in Anhui Province. *Comp. Econ. Soc. Syst.* **2013**, *6*, 133–146.
56. Xu, H.; Guo, Y.; Shi, S. Analysis on the impact of farmers' differentiation on Farmers' willingness to transfer Farmland—Estimation Based on structural equation model. *China Land Sci.* **2012**, *26*, 74–79. [CrossRef]
57. Liu, T.; Niu, L. Farmers' differentiation, land withdrawal intention and farmers' choice preference. *China Popul. Resour. Environ.* **2014**, *24*, 114–120.
58. Zou, W.; Wang, Z.; Zhang, B. Study on the impact of farmers' differentiation on Rural Homestead withdrawal behavior—Based on a survey of 1456 farmers in Jiangsu Province. *China Land Sci.* **2017**, *31*, 31–37.
59. Qian, L.; Qian, W.; Zheng, S. Citizenization ability, legal cognition and rural homestead transfer—Based on the investigation and demonstration of Wenzhou Experimental Area. *Agric. Econ. Issues* **2016**, *5*, 59–68.

sustainability MDPI

Article

Why Do Farmers Support Stable Land Ownership? Marketization with Chinese Characteristics

Ximing Zhang [1], Xiao Li [2,*] and Hui Wang [3]

[1] Land Consolidation Center of Shengzhou, Shaoxing 312000, China
[2] School of Economics and Management, Zhejiang University of Water Resources and Electric Power, Hangzhou 310018, China
[3] School of Public Affairs, Zhejiang University, Hangzhou 310058, China
* Correspondence: lix@zjweu.edu.cn; Tel.: +86-136-1654-7963

Abstract: Recent debates regarding marketization have focused on the relationship between the state and the market, while the grassroots and their everyday experiences have arguably been understudied. In this paper, we study marketization with the example of land marketization in China. Out of concern for the grassroots' perspective, we investigate Chinese farmers' perceptions with regard to stable land ownership of farmland, which are essential for land marketization in the backdrop of intensive land use conversion in China's urban periphery. Approximately 1200 farmer households were interviewed around 12 cities in mainland China. An ordered probit regression analysis of the survey results reveals a series of factors that explain the individual farmers' preference for stable land ownership. Among others, the decreasing size of farmer household and rural women's insecure property rights in farmland are identified as two grassroots-based characteristics underpinning China's ongoing transition to a more market-based farmland use institution. An important theoretical implication of our research is that the mainstream literature perhaps over-attributes China's marketization to the state and the market, while under-evaluating the spontaneous support from bottom-up.

Keywords: marketization; farmland conversion; grassroots; China

Citation: Zhang, X.; Li, X.; Wang, H. Why Do Farmers Support Stable Land Ownership? Marketization with Chinese Characteristics. *Sustainability* **2023**, *15*, 1729. https://doi.org/10.3390/su15021729

Academic Editor: Yu Cao

Received: 14 November 2022
Revised: 4 January 2023
Accepted: 7 January 2023
Published: 16 January 2023

1. Introduction

In the broad international development context, the last few decades have witnessed two different, albeit related, types of critical reflections on the global spread of marketization. The first kind of critique concentrates on the undesirable outcomes of market-oriented policies, mainly in terms of the intensification of socioeconomic inequalities and environmental degradation [1–3]. By comparison, the second group of scholars employ typically a Habermasian ideologiekritik approach to marketization, challenging its epistemological obsession with the market-state nexus, alongside its oversight of the grassroots [4–6].

The market-oriented reform in China offers a case to examine marketization against the background of economic globalization. The marketization of land, especially that of farmland, is one of the fundamental reforms in China's marketization reform. Farmland conversion in China reflects a variety of fundamental socioeconomic transitions in contemporary China and relates closely to the debate about whether the country has seen a sustained process of marketization since the reforms of the late 1970s [5–7]. On one hand, the reform of Household Responsibility System (HRS), characterized by the reallocation of farmland use rights to individual families, may be seen as a typical market-based "structural adjustment" [5–7]. On the other hand, the *Land Management Law* restricts almost all kinds of non-agricultural use of farmland to the eminent domain (with the exceptions of building town and village enterprises and village infrastructure), entitling local governments to convert farmland through land requisition [8].

In the literature on farmland conversion in China, there is also an intellectual divide regarding the global spread of marketization. Most existing research, arguably, seems to focus on the role of the state and the market in the marketization of land, often employing extensive analyses of the structure of farmland usufructs under the HRS and the various social and environmental issues related to the exercise of land requisition [9–11]. In contrast, a relatively small number of publications attempt to move beyond the state-market dichotomy to look at farmland conversion from a grassroots perspective that emphasizes the farmers' everyday cultural experiences [12–14].

In this paper, we continue to interrogate the dichotomist market-state discourse about farmland conversion in China, albeit through a statistical analysis in lieu of the qualitative approach that is more conventional in this kind of research. Based on a large set of survey data collected between 2008 and 2009 with 1209 farmer households across 12 localities in mainland China, our regression analysis suggests that rural grassroots opinions depend neither purely on the state nor the market. While a majority of the respondents appeared to prefer stable land ownership of farmland, their preference was driven by very specific localized factors, such as the shrinking size of farmer households due to local demographic transitions and female farmers' insecurity in terms of their post-marriage property rights in farmland. Given these findings, we argue that the marketization of China is not only attributable to the general state-market relationship, but also to some highly place-based local Chinese characteristics.

The remainder of our paper is organized as follows. The next two sections review the academic debate about China's marketization and its relation to the academic research on farmland conversion and stable land ownership in China's urban periphery. Following that is a presentation of the survey design and regression method employed in this study, with the model results reported and analyzed afterwards. Before the conclusion, the policy and academic implications of this research are discussed in detail.

2. Marketization, Developmental State, and Beyond

Recent reflections on marketization in the world primarily focus on the relation between the state and the market and, most inherently, epistemological structuralism [15,16]. These ideologiekritiks follow a quite different line of thinking from that of previous studies, which are primarily concerned with the substantive consequences of marketization policies, such as the intensification of socioeconomic inequalities and environmental degradations [17–20]. The fact that major international organizations nowadays have increased their funding support for Third-World-based nongovernmental organizations more or less reflects a determined departure from the old structuralist state-market dualism [21]. The role of the individual in marketization also needs to be emphasized. These are three main perspectives of marketization reform in different countries (see Table 1).

Table 1. Different perspectives on international development.

	Marketization	Developmental State	Grassroots' Agency
Philosophical foundation	structuralism	structuralism	agency & structure
Epistemological focus	How does the market influence the structure of civil society?	How does the state proactively regulate the market and influence the structure of the economy and civil society?	How does civil society interact with various constraining structures (including but not limited to the state and the market)?
Typical policy recommendations	free trade	regulation; trade protection;	place-based governance; local economic development
Major research subject	the market structures	the state institutions	the grassroots' perceptions and actions
Typical authors	Oi, J. [15]; Tong, Y. [16]	Cowling, K.; Tomlinson, P.R. [19]; Narins, T. [20]	Sulistyani, W [21]
Limitations	overemphasize the effect of the "invisible hand"(market) and ignore the role of state	believe that the state should regulate the market development	grassroots' forces cannot take effect without support from the state.

As a large developing country, China's market-oriented reform offers an example for the study of marketization against the background of globalization. The reform, which started in the late 1970s, was an institutional reform that aimed to transition from a planned economy to a market economy and establish a market economy system regulated by the state. The process of China's marketization has aroused much attention. Some studies are dedicated to measuring and accessing China's marketization levels with various aspects and indexes, including the proportion of economic resources allocated by the market, development of non-state-owned economy, government size, and so on [22–24]. These studies focus on the role of the developmental state and claim that state and government play the most important roles in the marketization process. The regulation of the state on the market has guided the market in the right direction, and the protection of trade order ensures a healthy market environment. Under China's communal land ownership system, land marketization depends more on state and government. Other studies pay attention to the influence of marketization on economic or social conditions, for example, on the income gap [25], economic growth [26,27], the implementation of social policy [28], social trust [29], etc. These market-centered theses demonstrate that marketization is the driving force of China's social development after the reform and opening-up. The allowance of free trade unleashed the dynamism of China's economic market and promoted social change. Allowing the free trade of land was the most important breakthrough. Land marketization plays a crucial role in the process of China's marketization. The major debate between these two perspectives is whether the state or the market dominates the process of land marketization in China. Further studies focus on the interaction between marketization and other factors, for instance urbanization, corruption, regional heterogeneity, decentralization, globalization, etc. [30–35]. These studies ignore the interaction between the state and market. Subsequently, some scholars focus on their interaction. Research on China's marketization also sees an overwhelming concentration on the role and interaction of the market and the state. In comparison, an attention to the grassroots is rare. While there are some studies that have extensively explored grassroots experiences and behaviors in marketization [23], the research on the opinion and perception of farmers about marketization is still rare, which may have influence on marketization.

The marketization of land, especially that of farmland, was one of the fundamental reforms in China's marketization reform. Land reform in China was initiated by the reform of the HRS in 1978, which reallocated the use-right of farmland to individual households and allowed the transaction of farmland. Then, a series of reforms, including the reform of farmland ownership confirmation and the "separation of three rights", guaranteed farmland to be freely transferable within agricultural use, while the transfer of farmland into nonagricultural use could only be achieved by land expropriation and land transfer of government. China's land reform has aroused much attention. Some scholars pay attention to the process and implication of land reform [36–38]; others focus on the impact of land reform on individual behaviors, for example the relation between land tenure reforms and land conservation investment [39]; others dedicate their studies to the effects of land reform on a macro level, for instance, on the scale of land use and rural development [40,41].

In the area of land reform, there is also an overwhelming concentration on the role and interaction of the market and the state. Land marketization reform in China, which originated with the reform of the household responsibility system in the late 1970s, was considered a state-guided reform from the very beginning [37]. The continued marketization of farmland is largely dependent on the reform of the state [38]. In the specific practice of land marketization, for example in farmland conversion, the state has a great impact on the degree of marketization [39,40]. For these reasons, land reform in China is portrayed by many scholars as a process of marketization that relies on the state's role in the market and is driven largely by government policies [42]. Despite the primary role that the state plays in land marketization, the market has also seen its growing importance in the process of land reform. For one thing, market capitalism is constantly promoted in the process of land marketization. For another, regulated as they are by the land market, land transactions are becoming increasingly common and normalized in rural areas, especially in the way of land lease.

3. Farmland Conversion and Stable Land Ownership in China

In China, marketization is one of the most conspicuous reforms since the late 1970s. It sounds to establish a market economy system regulated by the state, and to form a unified market operation mechanism and market system. In the 40 years since the reform and opening-up, one of the most important elements of China's economic and social development is the promotion of marketization with property rights reform. In particular, reform of rural land property rights is an important foundation for the marketization of land, which indicates the market allocation of land factors. With the ongoing reform, land ownership has become more stable and the transfer of rural land within agricultural use has become increasingly common.

Stable land ownership is essential for land transfer. In China, the stability of land ownership is reflected in many aspects, for example in land certificate, land adjustment, and land finance [43,44]. All of these are important components of land reform. It has been confirmed that stable land ownership helps to improve the scale and efficiency of land transfer by reducing transaction costs and improving land value [45,46]. With the continuous land reform over the years, the transactions of farmland within agricultural use has become more common in rural areas. However, the conversion of farmland into non-agricultural use is still strictly restricted by the state and can only be achieved by land requisition. According to *Land Management Law*, land requisition is the only legitimate way for farmland to be converted into non-agricultural use (with the exceptions for building town and village enterprises and village infrastructure). By land requisition, collectively owned land is transformed into state-owned land; then, once approval of land conversion has been attained, the original farmland can be used for non-agricultural use. After land requisition, farmers lose their land ownership permanently, which comprises a threat to the stable ownership of farmland.

In this sense, farmers' preference for stable land ownership indicates their longing for a smoother transaction of land use rights, and also their reluctance surrounding land requisition (instead of land conversion by market). Taken together, this reflects farmers' preference for land marketization.

There has been a large amount of research conducted on the marketization of China. Although recent decades have witnessed an increasing number of sophisticated and well-reasoned academic critiques regarding China's reform from both the angles of the state and the market [21,22,47], grassroots-based discourses remain relatively nascent.

This research gap is especially evident in studies about the marketization of land. Most of the literature seems to concentrate on the continued marketization of farmland following the household responsibility system reform in the late 1970s [48,49]. Another main research stream focuses on the role of the state in the marketization of land, mainly in the way of farmland conversion [50,51]. By comparison, discourses about the marketization of land from a farmer's perspective are on the minority side and, in terms of their research methods, rely overwhelmingly on ethnographic fieldwork [25–27]. Quantitative scholarship based on systematically-collected empirical data is a rarity. To fill this academic gap, we raise two essential research questions in our paper:

(a) What are the individual farmers' perceptions about the farmland usufruct under the HRS (Household Responsibility System)?
(b) Do farmers prefer more stable land ownership vis-à-vis the existing land requisition system to convert farmland into urban uses?

To address both questions systematically, we employ a rigorous social survey and statistical analysis approach to assess Chinese farmers' viewpoints on stable land ownership against the background of rapid farmland conversion in China's urban periphery.

4. Research Methods

4.1. The Sampling Framework

Between December 2008 and August 2009, we conducted a multistage geographical cluster sampling and interviewed 1209 farmers from different households around 12 cities (C1 to C12 in Figure 1) across east, north, south, and west China (see Figure 2 and Table 2). The survey was conducted in four major regions (Yangtze River Delta, Pearl River Delta, Yellow Sea and Bohai Sea, Chengdu Chongqing), which are located in east, south, north, and west China, respectively. According to the principle of stratified sampling, all cities of each region are classified into large, medium, and small ones, and one city is randomly selected from each category to generate 12 cities. In each city, two townships were randomly selected. Then from each township, we randomly chose five villages in the urban periphery, as the degree of marketization is higher in this place and the farmers have a stronger conception on marketization. Finally, about ten households were randomly selected in each village for one-to-one questionnaire interviews. Table 2 illustrates the sample distribution across the 12 cities, alongside the respondents' basic demographic information.

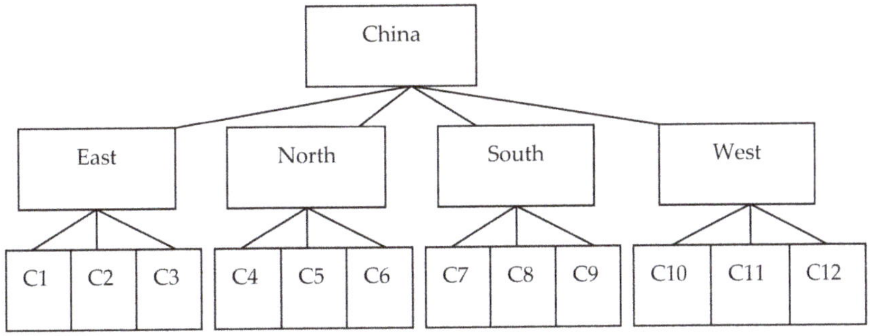

Figure 1. Multistage geographical clusters.

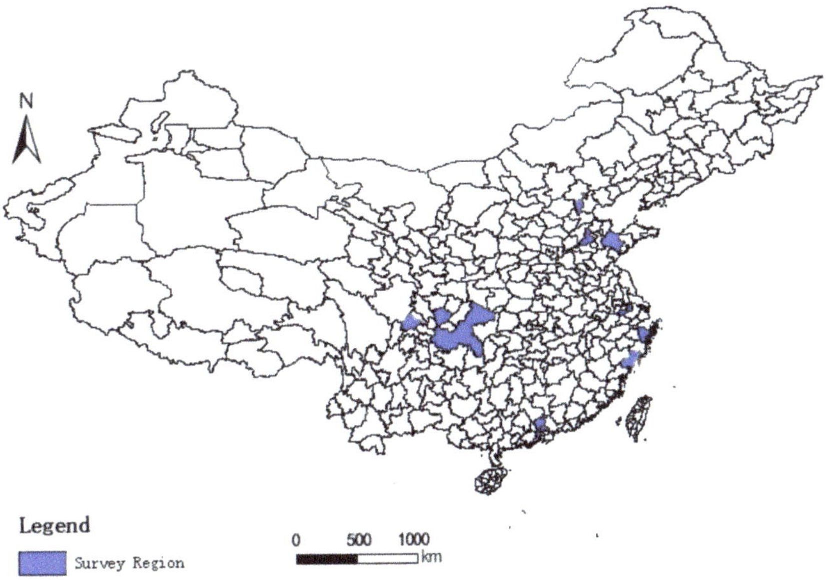

Figure 2. Distribution of the 12 survey cities.

Table 2. Sample distribution and interviewees' demographic information.

City	Sample Size	Male	Farmer	Ethnic	Party Member	Local Officials	Veteran
Yueqin (C1)	81	72.84%	96.3	0.00%	16.05%	4.94%	12.35%
Ninbo (C2)	102	78.43%	63.73	0.00%	40.20%	6.86%	12.75%
Jiangyin (C3)	101	80.2%	79.21	0.00%	8.91%	3.96%	8.91%
Sanhe (C4)	108	62.04%	87.04	0.00%	31.48%	7.41%	13.89%
Weifang (C5)	109	73.39%	84.4	0.00%	19.27%	4.59%	7.34%
Jinan (C6)	102	59.80%	95.1	0.00%	24.51%	9.80%	7.84%
Guangzhou (C7)	105	77.14%	42.86	0.00%	37.14%	3.85%	13.33%
Zhongshan (C8)	86	74.42%	5.81	0.00%	31.40%	5.81%	13.95%
Dongguan (C9)	90	76.67%	87.78	0.00%	16.67%	5.56%	11.11%
Chongqin (C10)	106	46.23%	85.85	0.00%	11.32%	6.60%	1.89%
Nanchong (C11)	108	66.67%	47.22	0.00%	11.11%	4.67%	2.78%
Chengdu (C12)	111	74.77%	73.87	0.9%	28.83%	7.21%	8.11%
Overall	1209	69.98%	71.05	0.08%	23.16%	5.97%	9.35%

4.2. The Questionnaire

Our survey questionnaire was structured with respect to three key interview questions (see Table 3), which were deployed collectively to measure the dependent variable, i.e., whether a respondent agrees with stable land ownership or not. Farmers' preference for stable land ownership is an indicator of their opinion on farmland marketization, as stable land ownership is necessary for the transaction of land use rights. Question 1 is an overarching, though abstract, query regarding an individual farmer's overall opinion about stable land ownership. In contrast, questions 2 and 3 draw on more specific economic interests related to the issue. The purpose of these two specific questions is to make sure farmers have good knowledge of the effects of stable land ownership and what it means to marketization. Question 2 reads as follows: "If the ownership of farmland is more stable, it would be more difficult to adjust each household's access to farmland based on the changing size of household under HRS. Do you think that is acceptable and appropriate?" In China, farmland was distributed equally among members of the collective and used to be adjusted according to demographic changes. Stable land ownership makes it more difficult to adjust land ownership according to family population growth, which means a loss for those households with expected population growth. This is the main reason that hinders farmers from agreeing to stable land ownership. Question 3 reads: "If the ownership of farmland is more stable, it can be transacted more freely in the market. Do you think that is acceptable and appropriate?" The free transaction of land is at the core of land marketization, and it has been confirmed that stable land ownership has a positive impact on rural land circulation by reducing transaction costs and improving land value [52,53]. If a respondent answered "yes" to questions 2 and 3 after saying "no" or "uncertain" to question 1, we would ask the person to revisit question 1. We would do the same if a farmer answered "no" to both questions 2 and 3 after saying "yes" or "uncertain" to question 1. Otherwise, we simply filed the original answer to question 1 as the result.

We also asked questions about each individual farmer's perception about land requisition because we assumed it to be an important factor in explaining whether a farmer would prefer stable land ownership or not. As well as this information, Table 4 lists all the explanatory variables we managed to collect data on through the survey.

4.3. The Regression

We applied an ordered probit regression method to analyze the survey results. Following Greene, we tested the respective probabilities of three responses with regard to an individual farmer's unobserved latent preference for stable land ownership.

$$\text{Prob}(y_i = -1 \mid x) = \prod(\alpha_1 - x\beta) \quad (1)$$

$$\text{Prob}(y_i = 0 \mid x) = \prod(\alpha_2 - x\beta) - \prod(\alpha_1 - x\beta) \quad (2)$$

$$\text{Prob}(y_i = 1 \mid x) = 1 - \prod(\alpha_2 - x\beta) \quad (3)$$

Equations (1)–(3) estimate, respectively, the probability for an unobserved farmer, i, to disagree, be uncertain, or agree with stable land ownership, given a matrix of empirical observations, x, and a vector of assessable coefficients, β. $\prod$ denotes the cumulative probability function corresponding to the standard normal distribution, while α_1 and α_2 stand for two threshold values or cut-off points on $\prod$ that are subject to estimation.

Table 3. Key survey questions regarding stable land ownership of farmland.

Question 1	Do you support more stable ownership of the collective farmland under the Household Responsibility System (HRS)?
Question 2	If the ownership of farmland is more stable, it would be more difficult to adjust each household's access to farmland based on the changing size of household under HRS. Do you think that is acceptable and appropriate?
Question 3	If the ownership of farmland is more stable, it can be transacted more freely in the market. Do you think that is acceptable and appropriate?
Measurement	−1 = No; 0 = Uncertain; 1 = Yes

Table 4. Supplementary questions in the survey.

Variable	Question and Measurement
Land Requisition (V1)	Has a land requisition taken place during 2004–2008? 1 = Yes; 0 = No
Dissatisfied with Requisition (V2)	Are you dissatisfied with the compensation for farmland requisition? 1 = Yes; 0 = Uncertain
Satisfied with Requisition (V3)	Are you satisfied with the compensation for farmland requisition? 1 = Yes; 0 = Uncertain
Land Reallocation (V4)	Have you ever experienced farmland reallocation under HRS? 1 = Yes; 0 = No
Age (V5)	Age in 2008
Gender (V6)	1 = Male; 0 = Female
Education (V7)	Years of full-time education
Household Size (V8)	Number of people per household in 2008
Change in Household Size (V9)	Has the size your household changed since the last round of farmland reallocation under HRS? 1 = Increased; 0 = Otherwise
Off-farm Income (V10)	Off-farm income in 2008/total household income in 2007
Per Capita Farmland in mu (V11)	Amount of arable land per capita within the household
Per Capita Income in RMB (V12)	Per capita income within the household
Party Member (V13)	Are there any members of the Chinese Community Party within the household? 1 = Yes; 0 = No
Local Official (V14)	Are there any local officials within the household? 1 = Yes; 0 = No
Veteran (V15)	Are there any veterans within the household? 1 = Yes; 0 = No
Working elsewhere (V16)	Are there any household members working elsewhere? 1 = Yes; 0 = No

We then specified two linear regression models to test the three probabilities mentioned above. In the first model, we analyzed all of the 1209 sample observations based on all of the explanatory variables included in Table 4, supplemented by 12 dummy control variables (i.e., C1 to C12), each corresponding to one of the 12 cities in Table 2. We followed the same approach in testing the second model, except that this time we only studied those who had experienced at least one farmland requisition between 2004 and 2008.

5. Empirical Findings

5.1. Descriptive Statistics

Table 5 reports the descriptive statistics regarding the dependent variable, i.e., whether an individual farmer supports stable land ownership or not. It is conceivable that a higher percentage (47.48% vs. 44.83%) of the farmers in our sample agreed to more stable land ownership of the farmland. Table 6 summarizes the descriptive statistics about the main independent variables, though it excludes the geographic control variables. As only 823 of the 1209 respondents had experienced farmland requisition between 2004 and 2008, the sample size for V2 and V3 shows only as 823 in Table 6.

Table 5. Descriptive statistics about farmers' preference for stable land ownership.

City	Cases	Do You Support Stable Land Ownership?					
		Agree	Percent	Disagree	Percent	Uncertain	Percent
Yueqin (C1)	81	34	41.98%	20	24.69%	27	33.33%
Ninbo (C2)	102	56	54.90%	38	37.25%	8	7.84%
Jiangyin (C3)	101	60	59.41%	40	39.60%	1	0.99%
Sanhe (C4)	108	55	50.93%	41	37.96%	12	11.11%
Weifang (C5)	109	39	35.78%	65	59.63%	5	4.59%
Jinan (C6)	102	41	40.20%	57	55.88%	4	3.92%
Guangzhou (C7)	105	43	40.95%	56	53.33%	6	5.71%
Zhongshan (C8)	86	50	58.14%	32	37.21%	4	4.65%
Dongguan (C9)	90	42	46.67%	46	51.11%	2	2.22%
Chongqin (C10)	106	45	42.45%	46	43.40%	15	14.15%
Nanchong (C11)	108	50	46.30%	52	48.15%	6	5.56%
Chengdu (C12)	111	59	53.15%	49	44.14%	3	2.70%
Overall	1209	574	47.48%	542	44.83%	93	7.69%

Table 6. Independent variables' descriptive statistics (sample size = 1209).

Variable	Cases	Mean	Sth	Min	Max
Land Requisition (V1)	1209	0.680728	0.466388	0	1
Dissatisfied with Requisition (V2)	823	0.307412	0.461702	0	1
Satisfied with Requisition (V3)	823	0.244228	0.42989	0	1
Land Reallocation (V4)	1209	0.315964	0.745462	0	8
Age (V5)	1209	50.08768	11.60222	18	87
Gender (V6)	1209	0.699752	0.458556	0	1
Education (V7)	1209	7.732837	3.273755	0	19
Household Size (V8)	1209	3.8933	1.423326	1	8
Change in Household Size (V9)	1209	0.047974	0.213799	0	1
Off-farm Income (V10)	1209	0.038958	0.195833	−3.42466	1
Per Capita Farmland in *mu* (V11)	1209	0.217085	0.512763	0	5
Per Capita Income in RMB (V12)	1209	15766.83	20578.99	−14350	241666.7
Party Member (V13)	1209	0.331679	0.471011	0	1
Local Official (V14)	1209	0.080232	0.271764	0	1
Veteran (V15)	1209	0.179487	0.383919	0	1
Working elsewhere (V16)	1209	0.630273	0.482931	0	1

Table 7 illustrates the local proportion of respondents who had experienced farmland requisition (i.e., those who report V1 = 1) across the 12 sample cities. It is conceivable that those cities that show a higher percentage of farmer respondents who had experienced land requisition also tend to display a higher proportion of farmers who support stable land ownership, for example, Ninbo (C2) and Jiangyin (C3) in east China.

Table 7. Local breakdown of respondents who experienced land requisition.

City	Cases	Cases V1 = 1	Proportion of Cases with V1 = 1
Yueqin(C1)	81	61	75.31%
Ninbo(C2)	102	87	85.29%
Jiangyin(C3)	101	80	79.21%
Sanhe(C4)	108	54	50.00%
Weifang(C5)	109	71	65.14%
Jinan(C6)	102	98	96.08%
Guangzhou(C7)	105	38	36.19%
Zhongshan(C8)	86	20	23.26%
Dongguan(C9)	90	35	38.89%
Chongqin(C10)	106	74	69.81%
Nanchong(C11)	108	95	87.96%
Chengdu(C12)	111	110	99.10%
Overall	1209	823	68.07%

This prompted us to focus on the 823 respondents who had experienced at least one farmland requisition between 2004 and 2008. Table 8 is the counterpart of Table 6 after we focused on the 823 farmers who had had their farmland expropriated. Table 9 then shows a local breakdown of the farmers' satisfaction level about the compensation they had received for the latest land requisitions.

Table 8. Independent variables' descriptive statistics (sample size = 823).

Variable	Cases	Mean	Std	Min	Max
Land Requisition (V1)	823	0.307412	0.461702	0	1
Dissatisfied with Requisition (V2)	823	0.244228	0.42989	0	1
Satisfied with Requisition (V3)	823	0.368165	0.81906	0	8
Land Reallocation (V4)	823	49.33779	11.90729	18	87
Age (V5)	823	0.712029	0.453093	0	1
Gender (V6)	823	7.880316	3.35935	0	19
Education (V7)	823	3.73147	1.422642	1	8
Household Size (V8)	823	0.053463	0.225092	0	1
Change in Household Size (V9)	823	0.027003	0.171028	−3.42466	1
Off-farm Income (V10)	823	0.112067	0.296879	0	3.19
Per Capita Farmland in *mu* (V11)	823	15,839.72	21,421.61	−1136	241,666.7
Per Capita Income in RMB (V12)	823	0.339004	0.473659	0	1
Party Member (V13)	823	0.072904	0.260137	0	1
Local Official (V14)	823	0.18955	0.392184	0	1
Veteran (V15)	823	0.643985	0.479111	0	1

5.2. Regression Results

In this section, we chose ordered probit regression as our study method. Ordered probit regression is a special method of ordinal regression that is suitable for data analysis where the dependent variable is an ordered categorical variable. The dependent variable of our study - farmers' preference for stable land ownership - is a categorical variable with three categories (agree, disagree, uncertain). The three categories have a certain order, which meets the requirements to use this method.

Table 9. Local breakdown of farmers' satisfaction with land requisition.

City	Obs	Satisfaction with Compensation for Land Requisition					
		Satisfied (V3 = 1)	Percent	Dissatisfied (V2 = 1)	Percent	Unsure (V2 = 0 = V3)	Percent
Yueqin(C1)	61	11	18.03%	40	65.57%	10	16.39%
Ninbo(C2)	87	29	33.33%	40	45.98%	18	20.69%
Jiangyin(C3)	80	39	48.75%	19	23.75%	22	27.50%
Sanhe(C4)	54	31	57.41%	15	27.78%	8	14.81%
Weifang(C5)	71	56	78.87%	2	2.82%	13	18.31%
Jinan(C6)	98	57	58.16%	26	26.53%	15	15.31%
Guangzhou(C7)	38	8	21.05%	12	31.58%	18	47.37%
Zhongshan(C8)	20	2	10.00%	6	30.00%	12	60.00%
Dongguan(C9)	35	4	11.43%	4	11.43%	27	77.14%
Chongqin(C10)	74	23	31.08%	29	39.19%	22	29.73%
Nanchong(C11)	95	51	53.68%	30	31.58%	14	14.74%
Chengdu(C12)	110	58	52.73%	30	27.27%	22	20.00%
Overall	823	369	44.84%	253	30.74%	201	24.42%

The results of two ordered probit regressions are presented in Table 10, with the standard errors parenthesized immediately below the corresponding coefficient estimates. The outcomes of the first regression suggest that those farmers who have experienced land requisitions (V1 = 1) tend to prefer stable land ownership. These results indicate that the loss of land ownership and of chance to participate in the land market caused by land requisition leads to a preference for stable land ownership. Experience of land reallocation (V4 = 1) had no significant effect. This may be because land reallocation is conducted within the village according to demographic change; while this leads to changes in land ownership, the changes are more expected than land requisition, thus leading to less worry about the stability of land ownership. Related to this, farmers who have not witnessed any increase in household size (V9 = 0) tended to prefer stable land ownership. The reason may be that, because farmland in China is equally distributed amongst members of the village collective and are adjusted according to change of population, those with decreased household size are more worried about the adjustment of land ownership. For demographic characteristics, the age (V5) and education (V7) of farmers has no significant effect, yet female farmers (V6 = 0) significantly preferred stable land ownership. That may be explained by the fact that women are more vulnerable in terms of land rights, as they are more likely to move due to marriage and other reasons, which makes it difficult for their land rights to be effectively guaranteed [54]. Family characteristics, household size (V8), off-farm income (V10), per capita farmland (V11), and per capita income (V12) have no significant influence, which may indicate that family characteristics are not strongly related to the stability of land ownership. In terms of social relations, there was no significant effect on those respondents whose families included veterans (V15 = 1) and members working elsewhere (V16 = 1); however, those whose families involve party members (V13 = 1) and local officials (V14 = 1) tended to disagree with stable land ownership, albeit at a lower level of statistical significance. In terms of geographical features, farmers from north China's cities of Weifang (C5 = 1) and Jinan (C6 = 1) appeared to disagree significantly with stable land ownership; this is in contrast with those from the city of Zhongshan (C8 = 1) in the south. These results may be explained by the fact that marketization in the south is of a higher degree than that in the north, and so farmers in the south have a stronger market conception that leads to a preference for stable land ownership

The second regression paid attention to attitudes toward compensation for land requisition, instead of experience of land requisition. It was found that farmers who were unsatisfied with compensation for land requisition (V2 = 1) tended to support stable land ownership. That may because those who were unsatisfied with compensation were more unwilling to undergo land requisition and instead prefered to hold land or to transfer land on market, which all rely on stable land ownership.

Table 10. Results of two ordered probit regressions.

Explanatory Variables	Regression Model 1	Regression Model 2
Land Requisition (V1)	0.472 ***	
	(0.0966)	
Dissatisfied with Requisition(V2)		0.249 **
		(0.108)
Satisfied with Requisition (V3)		0.0224
		(0.117)
Reallocation (V4)	0.0217	−0.0244
	(0.0538)	(0.0595)
Age (V5)	−0.000538	−0.00391
	(0.00381)	(0.00474)
Gender (V6)	−0.263 ***	−0.248 **
	(0.0803)	(0.0987)
Education (V7)	0.00287	0.00494
	(0.0124)	(0.0148)
Household Size (V8)	0.0222	0.0238
	(0.0286)	(0.0352)
Change in Household Size (V9)	−0.539 ***	−0.615 ***
	(0.175)	(0.195)
Off-Farm Income (V10)	0.165	0.229
	(0.134)	(0.162)
Per Capita Farmland (V11)	0.0495	0.0209
	(0.0864)	(0.165)
Per Capita Income (V12)	-1.22×10^{-6}	-2.23×10^{-6}
	(1.91×10^{-6})	(2.19×10^{-6})
Party Member (V13)	−0.143 *	−0.243 **
	(0.0836)	(0.0996)
Local Official (V14)	−0.263 *	−0.129
	(0.139)	(0.171)
Veteran (V15)	0.165	0.168
	(0.103)	(0.126)
Working Elsewhere (V16)	0.0258	0.104
	(0.0786)	(0.0979)
Weifang (C5)	−0.413 **	−0.0282
	(0.164)	(0.199)
Jinan (C6)	−0.503 ***	−0.306 *
	(0.170)	(0.182)
Zhongshan (C8)	0.405 **	0.750 **
	(0.189)	(0.379)
Cutoff point 1 (α_1)	−0.0691	−0.467
	(0.299)	(0.351)
Cutoff point 2 (α_2)	0.134	−0.231
	(0.300)	(0.352)
Log pseudo likelihood	−1060.1579	−734.54207
Wald chi^2	76.36(25)	48.32(26)
Prob > chi^2	0.0000	0.0050
Pseudo R^2	0.0371	0.0353
Observations	1209	823

Note: *, ** and *** stands for significant respectively at 10%, 5% and 1% level of confidence.

The results of other variables were generally consistent with those of the first regression, although the second test identified the change in household size (V9) as the most significant explanatory variable. For those farmers whose families had ceased to expand (V9 = 1), stable land ownership tended to be their strongly preferred policy option. In addition, the second regression also indicated that females (V6 = 0) tended to support stable land ownership, while those with party members in the families (V13 = 1) tended to disagree. Farmers from Zhongshan (C8 = 1) remained significantly more supportive of stable land ownership in contrast to those from Jinan (C6 = 1), although the latter showed less significance compared with the same figure in the first regression.

6. Conclusions and Discussion

6.1. Conclusions

In this research, we discussed the impact of several factors on whether individual farmers support stable land ownership or not. Our findings entailed a series of scientific and empirical implications, as follows:

(1) The farmers' dissatisfaction surrounding land requisition was a main driving factor in their support for stable land ownership. However, this proved not to be the primary reason. Rather, the explanatory variable regarding the change in household size (V9) turned out to have a larger magnitude as well as a higher level of significance than farmers' dissatisfaction with land requisition. Pervasive rural-to-urban migration, combined with the policy of family planning, has meant that many rural families in recent decades have seen a substantial decline in size compared with the multi-generational large farmer households that used to be very typical in the Chinese countryside. Under the HRS, collectively owned farmland is due to be reallocated based on household size every 15 to 30 years, which would clearly defy the interests of those shrinking families, who thus understandably would prefer stable land ownership to the HRS.

(2) Gender played an important role, according to our findings, in explaining female farmers' preference for stable land ownership. In China, a woman after a marriage is supposed to leave her own parents and move into her husband's household. This transition has farmland implications. For the female farmer's husband, the increase in household size usually entitles him (rather than his wife), on behalf of the household, to more farmland in the next round of reallocation under HRS. However for the female farmer's pre-marriage family, the decrease in household size may lead to a cut in their farmland quota. The issue is ultimately attributable to the traditional patriarchic economy in the Chinese countryside, where women tend to be more vulnerable in terms of their economic interests.

(3) The interaction between the state and the market is conceivable in farmers' day-to-day perceptions about land rights. Farmers are increasingly conscious of their land-related financial interests and seek fairer, market-based compensations to justify their farmland requisitions. The more dissatisfied they are with the compensations, the more strongly they prefer stable land ownership.

Finally, geography has a relatively minor impact on farmers' preferences for stable land ownership. Of the 12 cities, only Jinan (C6) and Zhongshan (C8) displayed modestly significant coefficient estimates. This demonstrates that farmers in different urban suburbs share similar preferences for stable land ownership. Combined with the descriptive statistics above, the proportion of "agree" and "disagree" regarding stable land ownership was mostly consistent, except for Zhongshan (C8) and Jiangyin (C3). The regression excluded geographic effects by controlling spatial fixed effects. Some research on farmers' perceptions of property rights security and land tenure stability [55,56], which takes different areas as samples, have reached similar conclusions. It can be proven that regional background has little effect on farmers' perception of land.

6.2. Further Discussion

From an intellectual perspective, this paper demonstrates a rigorous social survey and statistical analysis approach to systematically study farmers, as opposed to studying the state and the market, which are the two usual subject matters in this kind of research. If stable ownership of farmland can be considered a feature of market-based land use policies, our empirical research simply revealed some important grassroots-based local Chinese characteristics that have been driving the country's continuing marketization. As local governments are the actual managers of urban land, the existing literature on land marketization has mainly focused on the influence of local governments' behaviors on urban land marketization [57–59]. However, the ultimate goal of land marketization is to build a unified urban and rural land market. As such, present studies on the release of the market value of rural land and the promotion of the rural construction land market is insufficient. Our analysis focuses on farmers' perceptions of stable ownership and provides some evidence of their support for further marketization. Some studies have proven that clear ownership of rural collective land in some rural areas in China contributes to promoting the price of state-owned industrial land and improving the marketization level of urban and rural construction land [60]. Farmers' preference for stable land ownership will promote the identification of property rights, which benefits the formation of a rural land market.

In conclusion, we argue that the Chinese version of marketization, as reflected in the farmland domain, involves some unique grassroots-based local Chinese characteristics. As there is a large rural area in China, the study on Chinese land marketization cannot ignore farmers' perceptions. Those farmers who support stable land ownership tend to be from smaller households, female, and dissatisfied with land requisition. Their preference for stable land ownership is aligned with, though not always determined by, the marketization of the overall economy, of which the land market is an important element. We shall continue to orientate our future research to the farmer grassroots, hopefully supplemented our further study with more in-depth qualitative fieldwork and action research that would analyze the Chinese farmers' everyday actions in contemporary China's urban periphery.

Some researchers have mentioned that farmers' residential ownership and farmland ownership will affect their knowledge about, attitude towards, and practice of rural land changes. In contrast, our paper focused on the impact of land changes on land ownership preference. Farmers' knowledge and attitude will also affect their preferences. Thus, the interaction of these three factors is subject to further research. There is also an area of the literature that pays attention to the fragmentation of land ownership and discusses its evolution, content, and disadvantages to emphasize the importance of stable land ownership in sustainable land use [61]. This is worthy of further discussion, combined with our study.

However, there are some limitations to our study. For example, we used data collected between 2008 and 2009 for the empirical test, which was not up to date and may have reduced the effectiveness of our findings in explaining the reality. Furthermore, while we concentrated on China's marketization with the example of land marketization, more examples of cases are needed to gain a more comprehensive and profound knowledge of the role of grassroots in marketization.

Author Contributions: Conceptualization, X.Z. and H.W.; methodology, X.L.; validation, X.L. and H.W.; formal analysis, X.Z.; investigation, H.W.; data curation, H.W.; writing—original draft preparation, X.Z. and X.L.; writing—review and editing, H.W.; supervision, X.L. All authors have read and agreed to the published version of the manuscript.

Funding: This research was funded by the National Social Science Major Foundation of China, grant number 15ZDA024; National Natural Science Foundation of China, grant number 21BGL163.

Informed Consent Statement: Informed consent was obtained from all subjects involved in the study.

Data Availability Statement: The data presented in this study are available on request from the corresponding author. The data are not publicly available due to privacy.

Conflicts of Interest: The authors declare no conflict of interest.

References

1. Liu, T.; Cao, G.; Yan, Y.; Wang, R.Y. Urban land marketization in China: Central policy, local initiative, and market mechanism. *Land Use Policy* **2016**, *57*, 265–276. [CrossRef]
2. Jiang, R.; Lin, G. Placing China's land marketization: The state, market, and the changing geography of land use in Chinese cities. *Land Use Policy* **2021**, *103*, 105293. [CrossRef]
3. Wang, F.; Liu, Y. How unique is 'China Model': A review of theoretical perspectives on China's urbanization in anglophone literature. *Chin. Geogr. Sci.* **2015**, *1*, 15. [CrossRef]
4. Wang, R.; Tan, R. Efficiency and distribution of rural construction land marketization in contemporary China. *China Econ. Rev.* **2020**, *60*, 101223. [CrossRef]
5. Konadu-Agyemang, K.; Newman, K.; Mohan, G.; Brown, E.; Milward, B.; Zack-Williams, A.B. Structural adjustment: Theory, practice and impacts. *Econ. Geogr.* **2000**, *78*, 245. [CrossRef]
6. Meng, G. The Household Responsibility System, Karl Marx's theory of property and Antony M. Honore's concept of ownership. *Sci. Soc.* **2019**, *83*, 300–326. [CrossRef]
7. Stig, T. Organizing rural China—Rural China organizing. *China Q.* **2013**, *213*, 212–214.
8. Shen, X.; Wang, L.; Wu, C.; Lv, T.; Lu, Z.; Luo, W.; Li, G. Local interests or centralized targets? How China's local government implements the farmland policy of Requisition–Compensation Balance. *Land Use Policy* **2017**, *67*, 716–724. [CrossRef]
9. Ren, X.; Yu, J. China's policy learning during the reform and opening period: The transformation of urban land-use policy. *China Public Adm. Rev.* **2016**, *2*, 37–49.
10. Zhang, Y. From state to market: Private participation in China's urban infrastructure sectors, 1992–2008. *World Dev.* **2014**, *64*, 473–486. [CrossRef]
11. Su, L.; Tang, J.; Qiu, H. Intended and unintended environmental consequences of grassland rental in pastoral China. *J. Environ. Manag.* **2021**, *285*, 112–126. [CrossRef]
12. Chen, Q.; Cai, Y.; Liu, F.; Zhou, Q.; Zhang, H. Farmers' perception to farmland conversion: A questionnaire survey in Xining City, Qinghai Province, China. *Chin. Geogr. Sci.* **2013**, *5*, 13. [CrossRef]
13. Zhu, D. Farmers are growing further and further from the land: Land transfer and the practice of three rights separation in China. *Soc. Sci. China* **2021**, *42*, 24–43.
14. Zhou, B.; Zhou, S.; Lv, L.; Lu, C. Interprovincial variations of farmland expropriation in China and the estimation of lost-land farmers. *Chin. Agric. Sci. Bull.* **2015**, *31*, 284–290.
15. Oi, J.C. The Role of thelLocal state in China's transitional economy. *China Q.* **1995**, *144*, 1132–1149. [CrossRef]
16. Tong, Y. The imbalance between government regulation and control and market adjustment in China. *Reform. Strategy* **2014**, *14*, 459.
17. Li, O.Z.; Su, X.; Yang, Z. State control, access to capital and firm performance. *China J. Account. Res.* **2012**, *2*, 25.
18. Shadare, G.A. The governance of Nigeria's social protection: The burdens of developmental welfarism? *Societies* **2022**, *12*, 20. [CrossRef]
19. Cowling, K.; Tomlinson, P.R. Post the 'Washington Consensus': Economic governance and industrial strategies for the twenty-first century. *Camb. J. Econ.* **2011**, *35*, 831–852. [CrossRef]
20. Narins, T. The China triangle: Latin America's China boom and the fate of the Washington Consensus. *AAG Rev. Books* **2017**, *5*, 17–19. [CrossRef]
21. Sulistyani, W. *Encountering Development: The Making and Unmaking of the Third World*; Princeton University Press: Princeton, NJ, USA, 1995.
22. Dong, X.; Hao, L. A Quantitative Study of China's Marketization Process: The Marketization Index of Last 30 Years. *Contemp. Econ. Manag.* **2010**, *32*, 6.
23. Tian, W.M. An Empirical Analysis of the Impact of China's Marketization Process on Income Distribution. *Contemp. Financ. Econ.* **2012**, *448*, 259–270.
24. Pan, Z. New Situation and Countermeasures of China's Marketization Process: Based on the Measurement of the Marketization Index of 2006—2013. *J. Zhejiang Shuren Univ. (Humanit. Soc. Sci.)* **2016**, *16*, 42–48.
25. Ping, W. China's Economic Marketization Process and the Study of the Measuring Indicator Setup. *Mod. Financ. Econ.* **2002**, *22*, 4.
26. Guan, X. China's Social Policy: Reform and Development in the Context of Marketization and Globalization. *Soc. Policy Adm.* **2000**, *34*, 115–130. [CrossRef]
27. Niu, L.; Accounting, S.O. Marketization, Auditor Industry Specialization and Earnings Quality. *J. Postgrad. Zhongnan Univ. Econ. Law* **2013**.
28. Fan, G.; Wang, X.; Zhang, L.W.; Zhu, H. Marketization Index for China's Provinces. *Econ. Res. J.* **2003**, *3*, 9–18.
29. Wu, F. China's recent urban development in the process of land and housing marketization and economic globalization. *Habitat Int.* **2001**, *25*, 273–289. [CrossRef]
30. Fan, G.; Wang, X.; Ma, G. The Contribution of Marketization to China's Economic Growth. *Econ. Res. J.* **2011**, *7*, 4–14.

31. Zhu, J. Urban Development under Ambiguous Property Rights: A Case of China's Transition Economy. *Int. J. Urban Reg. Res.* **2002**, *26*, 41–57. [CrossRef]
32. Jacobsen, G.D.; Office, E. Market-based policies, public opinion, and information. *Econ. Lett.* **2020**, *189*, 109018. [CrossRef]
33. Drechsler, M. Impacts of human behavior in agri-environmental policies: How adequate is homo oeconomicus in the design of market-based conservation instruments? *Ecol. Econ.* **2021**, *184*, 107002. [CrossRef]
34. Wei, Y. Decentralization, Marketization and Globalization: The Triple Process Underlying Regional Development in China. *Asian Geogr.* **2001**, *20*, 7–23. [CrossRef]
35. Bezabih, M.; Holden, S.T.; Mannberg, A. The role of land certification in reducing gaps in productivity between male-and female-owned farms in rural Ethiopia. *J. Dev. Stud.* **2016**, *52*, 360–376. [CrossRef]
36. Huang, Z.H.; Xue-Jun, D.U. Review, Comparison and Implications of the 60 Years' Land Reforms in Chinese Mainland and Taiwan. *China Land Sci.* **2010**, *4*, 5.
37. Schwarzwalder, B.; Prosterman, R.; Jianping, Y.; Riedinger, J. An update on China's rural land tenure reforms: Analysis and recommendations based on a seventeen-province survey. *Columbia J. Asian Law* **2002**, *16*, 141.
38. Zhou, Y.; Li, X.; Liu, Y. Rural land system reforms in China: History, issues, measures and prospects. *Land Use Policy* **2019**, *91*, 104330. [CrossRef]
39. Wesseler, J.H.H.; Heerink, N.; Qu, F. Land tenure reforms and land conservation investments in China—What does real option value theory tell us. *Rev. Econ. Financ.* **2013**, *3*, 19–33.
40. Zhu, Q.; Lu, L.; Fang, H. The driving forces of the campaign of new socialist countryside construction—On the land ownership reforms in the rural areas. *J. China Agric. Univ. (Soc. Sci. Ed.)* **2006**, *1*, 5.
41. Liu, L.; Chen, Z.; Chen, Y. Effects and mechanism of market-oriented land reforms on the scale of industrial land use:an empirical study on 46 cities in China. *Prog. Geogr.* **2015**, *42*, 233–243.
42. Wang, X.; Hu, S. In the spirit of radical liberalism: A historical review of land reforms in China from the 1970s to today. *Camb. J. Econ.* **2022**, *3*, 3. [CrossRef]
43. Chen, Z.; Zhang, Y. Has rural finance promoted the agriculture large-scale operation in China:Based on the survey data of two national family farm demonstration bases. *Issues Agric. Econ.* **2022**, *5*, 83–97.
44. Qian, L.; Feng, Y.; Lu, H. Impact of land tenure stability on farmer's cultivated land quality protection behavior: Analysis based on adjustment effect of the new round land certification. *J. Nanjing Agric. Univ. (Soc. Sci. Ed.)* **2021**, *21*, 104–115.
45. Zhao, J.; Barry, P. Effects of credit constraints on rural household technical efficiency: Evidence from a city in northern China. *China Agric. Econ. Rev.* **2014**, *6*, 654–668. [CrossRef]
46. Zhou, C.; Zhang, H.; Lyu, K.; Zhang, C. The Influence of land rights stability on soil quality from the perspective of transfer contract arrangement. *Chin. J. Agric. Resour. Reg. Plan.* **2022**, *43*, 237–247.
47. Wu, F. *Retreat from a Totalitarian Society: China's Urbanism in the Making*; Wiley-Blackwell: Oxford, UK, 2011; pp. 701–712.
48. Giddens, A. Central problems in social theory: Action, structure and contradiction in social analysis. *Am. J. Sociol.* **1980**, *74*, 188–189.
49. Huang, Z.; Wei, Y.D.; He, C.; Li, H. Urban land expansion under economic transition in China: A multi-level modeling analysis. *Habitat Int.* **2015**, *47*, 69–82. [CrossRef]
50. Lin, Z. Vertical decentralization and marketization in state-owned basic industries: Evidence from the state-driven reform of the telecommunications industry. *China Public Adm. Rev.* **2016**, *1*, 42–56.
51. Chen, B.; Cao, W. From opportunity equity to income equality: Income distribution dynamics in China. *Comp. Econ. Soc. Syst.* **2013**, *6*, 44–59.
52. Xia, M. *The Dual Developmental State: Development Strategy and Institutional Arrangements for China's Transition*; Ashgate Pub Ltd.: Farnham, UK, 2017.
53. Denise, H.L.; Yang, D. Englander, "Land Management in Rural China and its Gender Implications". *Fem. Econ.* **2007**, *13*, 35–61.
54. Lu, S.; Wang, H. Market-oriented reform and land use efficiency: Evidence from a regression discontinuity design. *Land Use Policy* **2022**, *115*, 106006. [CrossRef]
55. Qian, L.; Feng, Y.; Lu, H.; Chen, H. The Influence of Property Right Security Perception on Farmers' Farmland Quality Protection Behaviors: Taking Guangxi as an Example. *China Land Sci.* **2019**, *33*, 93–101.
56. Li, N.; Tang, L.; Che, X.; Shi, X.; Ma, X. Does the democratization level of village governance affect perceptions of security and integrity of land rights? -An analysis from the perspective of social network abundance. *J. Rural Study* **2022**, *94*, 305–318. [CrossRef]
57. Mu, Y.; Qian, Z. Will the Marketization Level of Primary Land Market be Upgraded by Land Financial Dependence?:Based on the Test of a Provincial-Level Panel Data in China from 2003 to 2015. *China Land Sci.* **2018**, *32*, 8–13.
58. Zhou, L.; Fan, J.; Yu, X. The Bilateral Effect of Intergovernmental Competition on the Level of Urban Land Marketization: Based on the Different Functions of Fiscal Competition and Investment Attraction Competition. *China Land Sci.* **2019**, *33*, 60–68.
59. Fan, X.; Qiu, S.N.; Sun, Y.K. Land finance dependence and urban land marketization in China: The perspective of strategic choice of local governments on land transfer. *Land Use Policy* **2020**, *99*, 105023. [CrossRef]

60. Huang, Z.; Du, X.J. Does the Marketization of Collective-owned Construction Land Affect the Integrated Urban-Rural Construction Land Market? An Empirical Research Based on Micro-level Land Transaction Data in Deqing County, Zhejiang Province. *China Land Sci.* **2020**, *34*, 18–26.
61. Zhu, D.L.; Wang, J.; Lin, R.R. Outlook for China's rural collective land system reform: Reviews from "Roundtable Forum for Land Policy and Law 2014". *China Land Sci.* **2014**, *28*, 89–94.

sustainability

Article

The Impact of Farmers' Perception on Their Cultivated Land Quality Protection Behavior: A Case Study of Ningbo, China

Xiaoying Wang, Hangang Hu, Aifeng Ning *, Guan Li and Xueqi Wang

Law School, Ningbo University, Ningbo 315211, China; wangxiaoyingzj@126.com (X.W.); huhangang@nbu.edu.cn (H.H.); liguan@nbu.edu.cn (G.L.); wangxueqi@nbu.edu.cn (X.W.)

* Correspondence: ningaifeng@nbu.edu.cn

Abstract: Farmers' protection behavior largely depends on their perceived value of cultivated land quality protection. However, existing research shows that the impact path of these perceived factors on farmers' cultivated land protection behavior is not clear. Based on the survey data of 288 farmers in Ningbo City, this study empirically analyzed the impact of farmers' perception on their cultivated land quality protection behavior through structural equation modeling (SEM). The results showed that farmers' cultivated land quality protection behavior largely depended on perceived value, and they followed the logic paradigm of "perceived tradeoff→perceived value→behavioral intention→behavioral response". Among them, farmers' perceived value comes from farmers' comprehensive tradeoff of benefits and risks in the process of cultivated land quality protection. In other words, improving farmers' perceived benefits and reducing perceived risks is conducive to improving farmers' perceived value of cultivated land quality protection. The above findings are helpful to improve farmers' behavior of farmland land quality protection and provide new ideas and empirical basis for the design and improvement of cultivated land quality protection policies.

Keywords: cultivated land quality protection; farmer's behavior; perceived value; structural equation model (SEM)

Citation: Wang, X.; Hu, H.; Ning, A.; Li, G.; Wang, X. The Impact of Farmers' Perception on Their Cultivated Land Quality Protection Behavior: A Case Study of Ningbo, China. *Sustainability* **2022**, *14*, 6357. https://doi.org/10.3390/su14106357

Academic Editor: Andres Seco Meneses

Received: 30 March 2022
Accepted: 20 May 2022
Published: 23 May 2022

1. Introduction

Cultivated land is the essential resource and condition for human survival, the basis of national economic development and social stability, and the fundamental guarantee for realizing people's food security [1–4]. The quantity and quality of cultivated land are closely related to sustainable human development [5–7]. Governments worldwide have long attached great importance to farmland protection [8–10]. However, the vast population and the excessive expansion of cities have caused enormous demand pressure on limited cultivated land resources [11,12]. Cultivated land resources have been continuously reduced, and many high-quality cultivated land areas have been occupied [9,13]. The contradiction among food security, economic development, and cultivated land resource protection have become increasingly prominent [14,15]. In the process of meeting the needs of rapid population and urbanization for grain size and structure, long-term problems, such as the excessive application of chemical fertilizers, insufficient organic recycling, and unreasonable farming methods, have emerged, thus leading to the deterioration of cultivated land quality [16].

To this end, countries worldwide have made many efforts to actively explore practical ways to protect arable land [17]. On a global scale, farmland protection is usually implemented through government intervention [18], mainly including demarcating agricultural protected areas through planning means [19], implementing compulsory agricultural land protection policies through legal means [20], and promoting agricultural land development rights trading through market means [21,22]. However, the protection of cultivated land quality is a systematic project. The government's participation alone is not enough to solve the problem [19].

As the direct users of cultivated land, farmers play an essential role in protecting such land and significantly improving its quality [23]. They are the last barrier to protecting arable land quality through the quality of cultivated land protection and soil improvement. Previous studies indicated that farmers' land-use behavior, including the choice of land-use type, plantation structure [24], land-related inputs [25], the use of fertilizers and pesticides [26], and agricultural waste resource utilization [27], have significant impacts on the cultivated land quality [28]. However, in practice, the consciousness of farmers about the quality of cultivated land protection policies and measures is generally not high [29]. Therefore, effectively improving farmers' enthusiasm in assuming their role as protectors of cultivated land quality is critical.

At present, the research on farmland protection from farmers' perspectives has achieved fruitful results, mainly focusing on the empirical analysis of demographic characteristics, socioeconomic characteristics, and institutional environment characteristics [30,31]. Recently, some scholars began paying attention to the influence of psychological factors at the microlevel on farmers' willingness and behavior toward farmland quality protection [32]. However, the explanations for farmers' behavioral decisions on farmland quality protection from the perspective of farmers' perception are limited, and only a few studies have taken this as one of many observed variables [33]. According to the theory of a farmer's behavior, the perceived value is the most direct cause of the formation of a farmer's behavioral attitude [34,35]. In other words, farmers' protection behavior largely depends on their perceived value of cultivated land quality protection [36].

In addition, perceived value also includes two different dimensions, perceived benefit and perceived risk. Compared with perceived benefits, perceived risk is also an essential factor affecting farmers' cultivated land protection behavior. Risk perception is the perceived probability that a specific (environmental) phenomenon (risk) will have negative consequences for individuals and society [37], and it is influenced by the subjective interpretation of risk bearers. Empirical studies showed that risk aversion significantly impacts pesticide use in Yunnan, China [38]. Therefore, one should carefully evaluate and deal with farmers' risk perception by applying appropriate strategies [39]. Farmers' perception of risk will directly impact farmland protection behavior, and agricultural production is usually characterized by significant risk and considerable government intervention [40]. As farmers are the direct subject of the implementation of cultivated land quality protection, the design of cultivated land quality protection policies should closely focus on farmers' behavior characteristics and the laws protecting cultivated land quality [41]. If the policy mechanism cannot conform to the motivation of farmers in the protection of cultivated land quality, the effect of the policy will be significantly reduced [42]. However, previous studies focused more on the promotion effect of perceived benefits on farmers' cultivated land protection behavior while ignoring the potential impact of perceived risks [43].

Given the weak links in the research into farmers' cultivated land quality protection behavior and the fact that the existing policy guidance is weak, in the current study, we take the theory of perceived value as a basis and include two antecedent variables of perceived value, namely perceived value benefit and perceived risk. Because of a high level of urbanization and industrialization in China, as well as the presence of relatively rich types of farmers still engaged in traditional agricultural production activities at the present stage, such as pure agricultural farmers, agriculture-oriented and part-time farmers, and pure migrant farmers, Ningbo city was taken as a case study. Heterogeneous farmers have significant differences in perceived value of cultivated land quality protection, which also shows an inconsistent influence path on cultivated land quality protection behavior. Therefore, the study on the effect path and influence degree of farmers' perceived value on their cultivated land quality protection behavior in developed areas will have wider application value for the formulation of targeted cultivated land quality protection policies for other developing countries and regions.

On the basis of survey data of 288 farmers in Ningbo City and the establishment of a structural equation model, in this study, we comprehensively clarify the psychological mechanism and behavioral logic behind farmers' cultivated land quality protection to provide a reference for improving the policy of farmland quality protection and standardizing farmers' cultivated land quality protection behavior. The aims of this study were as follows: (1) to build an overall model of farmers' cultivated land quality protection decision on the basis of benefit–risk balance; (2) to measure farmers' perceived benefits and risks of farmland quality protection; (3) to determine the influence of identifying perceived value on farmers' cultivated land quality protection behavior, with Ningbo, China as an example; and (4) to promote some suggestions on the construction of farmland protection hospitals and implementation of behaviors.

2. Theoretical Analysis and Research Hypothesis

2.1. Theoretical Analysis

Perceived value theory originated from the research on customer willingness and behavior in the field of product marketing. Zaithaml proposed that perceived value is based on the perspective of individual cognition, from the perspective of individual experience, the interests of a specific commodity, service or behavior [44]. The subjective comprehensive evaluation was formed by the trade-off comparison with the effort. Regarding the formation mechanism of perceived value, the "trade-off model" believes that perceived value depends on the individual's trade-off between the corresponding relationship between the benefits obtained and the costs paid. When the perceived benefits are higher than the perceived losses, the individual's perceived value level will be higher the more obvious its behavioral tendency is, and the possibility of actual behaviors will be greater [45,46]. It can be seen from the above analysis that the theory of perceived value clarifies the logic mechanism of the individual behavior decision-making process, that is, "cognitive weighing→perceived value→behavior willingness→behavior performance".

In the study of farmers' economic behavior, perceived value is also considered to be an essential basis for behavior [47]. Farmer behavior theory, focusing on the attitude of farmers, is the primary factor that affects farmers' behavior intention [34,40]. Perceived value is the most direct driver of peasant households' behavioral attitude as the "rational economic individual", and the microscopic operators of agricultural production continuously pursue the most significant benefits with the minimum cost [26]. This provides theoretical support for the research on farmers' behavior from the perspective of perceived value.

In fact, in the behavioral decision-making process for cultivated land quality protection, farmers also have a relatively comprehensive perception. Their perceived value is a subjective evaluation formed after weighing and comparing the perceived benefits and risks of cultivated land quality protection on the basis of their resource endowment and livelihood strategies [34,36,48]. When the perceived benefit is higher and the perceived risk is lower, their perceived value will increase [46]. Generally speaking, a higher perceived value of an individual toward a certain behavior stimulates a greater psychological intention. The willingness to engage in the behavior will also be higher. In other words, a higher perceived benefit of farmers' cultivated land protection behavior leads to higher cultivated land quality protection behavior [26]. Individual behavior is dominated by perceived value and is also affected by perceived level (the source of perceived value) [46]. The difference of individual's perceived level of a certain thing determines its different behavioral responses.

2.1.1. Factors Influencing Farmers' Perceived Value of Cultivated Land Quality Protection

As far as cultivated land quality protection is concerned, the production practice of cultivated land quality protection may produce many benefits. Among them, the perception of economic benefits is the most intuitive feeling of farmers on farmland input and output [31]. With the progress of industrialization and urbanization, cultivated land utilization has changed. However, the economic benefits such as increased yield and planting income are still the most intuitive feelings of farmers [32]. As human civilization entered the stage of ecological civilization, farmers gradually began to pay attention to the externalities brought by cultivated land protection, such as soil and water conservation, climate improvement, biodiversity protection, national food security, agricultural product security, and other ecological and social benefits [1,34,49]. All the above externalities of cultivated land protection can be captured by the brain of farmers to form the perception of benefits and evaluate possible benefits and their expectations of cultivated land protection behavior [21]. When perceived benefits can be generated and meet their expectations, their perceived value will increase [50]. Perceived risk also plays a decisive role in farmers' decision-making processes for farmland quality protection. Scott (1976) proposed that farmers in most developing countries take "safety first" as the principle of production and life. Under the survival ethics of "safety first", farmers do not pursue income maximization as they seek low risk distribution and high survival guarantee [51]. Farmers are very risk averse in agricultural production, and the improved family life brought by a higher-than-expected income is not enough to compensate for the devastating impact of a lower-than-expected income on families [52].

Farmers' perceived risk of cultivated land quality protection mainly comes from factor cost, inner anxiety, and worry about uncertainty [53]. Liu (2013) found a close connection between perceived risk and perceived value [54]. Among all factors influencing perceived value, perceived risk has a negative effect [55]. Therefore, higher perceived risk of farmers results in greater loss or concern related to cultivated land quality protection; hence, their perceived value of cultivated land quality protection is reduced. Therefore, this study proposes Hypothesis H1 and H2.

Hypothesis 1 (H1). *Farmers' perceived benefits in cultivated land quality protection positively impact their perceived value.*

Hypothesis 2 (H2). *Farmers' perceived risk of cultivated land quality protection has a negative impact on their perceived value.*

2.1.2. Impact of Farmers' Perception on Their Willingness to Protect Cultivated Land Quality

Protecting cultivated land quality refers to farmers' psychological intention to protect cultivated land quality. Generally, a significant positive correlation exists between an individual's perceived value and behavioral intention. This conclusion has been confirmed by many research results [36]. Studies on farmers' economic behaviors show that a higher perceived value level of behavior stimulates greater psychological intention and a higher willingness to engage in the behavior [26]. In addition, as rational economic individuals, farmers' behavioral intentions are driven by benefits [50]. Khamfeua and Toshiyuki (2012) studied the relationship between farmers' perceived benefits and their willingness to protect surrounding national reserves in Laos, and they found that farmers' perceived benefits of protection had a positive and significant relationship with their willingness to protect [56]. Moreover, rational individuals tend to increase returns and avoid risks [40].

The fundamental reason for farmers' implementing cultivated land quality protection is the value that farmers can perceive through the total balance between the perceived benefits and risks of cultivated land quality protection. If the expected benefits are greater than the expected costs, then the perceived benefits are higher than the perceived risks; a higher perceived value of cultivated land quality protection results in a stronger willingness to implement cultivated land quality protection. Therefore, Hypotheses H3, H4, and H5 are proposed in this study.

Hypotheses 3 (H3). *Farmers' perceived benefits in cultivated land quality protection positively impact their willingness to protect.*

Hypotheses 4 (H4). *Farmers' perceived risk of cultivated land quality protection has a negative impact on their willingness to protect.*

Hypotheses 5 (H5). *Farmers' perceived value of cultivated land quality protection positively impacts their willingness to protect.*

2.1.3. Impact of Farmers' Perception on Their Cultivated Land Quality Protection Behavior

Cultivated land quality protection behavior refers to the production behaviors taken by farmers in the process of agricultural production to maintain or improve the quality of cultivated land [50]. Such behavior includes returning straw to the field, planting green fertilizer, applying farm manure, using commercial organic fertilizer, testing soil formula fertilization, and other measures to improve barren cultivated land. These measures include those for improving the soil's ability to retain water, soil, and fertilizer; some examples are canal repair, land leveling, and deep plowing [40,48,52]. In this study, farmers' perceived value, its two antecedents (perceived benefits and perceived risks), and willingness to protect cultivated land quality jointly affect cultivated land quality protection behaviors. Wang and Guo (2020) found that perceived benefits positively impact farmers' cultivated land quality protection behavior when discussing the influence of perceived benefits and social networks on farmers' cultivated land quality protection behavior [33]. However, when farmers make accurate decisions, perceived risks significantly impact farmers' green agricultural production behaviors [26]. In addition, relevant studies have found that individual behavioral decisions result from perceived values formed after weighing and comparing perceived benefits and risks [57]. Therefore, in the decision-making process for farmland quality protection, if the expected protection benefits of farmers are more significant than the expected cost, then the farmers' perceived value of farmland quality protection will be high, and the condition becomes conducive to the implementation of farmland quality protection behavior by farmers. As a prevariable affecting the generation of specific behaviors, behavioral intention is shown as the possibility or tendency of individuals to choose specific behaviors and thus plays a vital role in predicting the generation of behaviors [36]. Behavioral intention reflects the degree to which an individual is willing to pay when choosing a particular behavior. A stronger individual's behavioral intention indicates a higher possibility of taking practical actions [58]. Similarly, in the decision-making process for farmland quality protection, a stronger farmers' behavioral intention toward farmland quality protection results in a greater probability of farmland quality protection in production practice. Therefore, we propose Hypotheses H6–H9.

Hypothesis 6 (H6). *Farmers' perceived benefits in cultivated land quality protection positively impact their behavior.*

Hypothesis 7 (H7). *Farmers' perceived risk of cultivated land quality protection has a negative impact on their protection behavior.*

Hypothesis 8 (H8). *Farmers' perceived value of cultivated land quality protection positively impacts their behavior.*

Hypothesis 9 (H9). *Farmers' willingness to protect cultivated land quality positively impacts their protection behavior.*

2.2. Model Design

Behavioral economics points out that behavioral decisions result from individuals' comparison of benefits and risks. Rational individuals always pursue maximum benefits with minimum risks, thus providing a theoretical basis for studying farmers' behavior of farmland investment from the perspective of perceived value theory. The "overall decision-making model of farmers' cultivated land quality protection behavior" was constructed (Figure 1). As shown in Figure 1, the overall decision-making model of farmers' cultivated land quality protection includes perceived benefits (PB), perceived risks (PR), perceived value (PV), behavior intentions (BI), and behavior responses (BR). The causal path relationship constitutes the internal logical mechanism of peasant household's behavior decision making for cultivated land quality protection. To reflect the overall composition of farmers' perception, we use the research results of previous studies on the multifunctional value of cultivated land to measure perceived benefits from three dimensions: perceived economic benefits (PEB), perceived ecological benefits (PECB), and perceived social benefits (PSB) [59,60]. Combined with the definitions of perceived risk by Sanjeev and Kenneth (2001), Zander et al. (2019), and Li et al. (2020), perceived risk herein is measured from three dimensions: perceived economic risk (PER), perceived psychological risk (PPR), and perceived situational risk (PSR) [26,41,55].

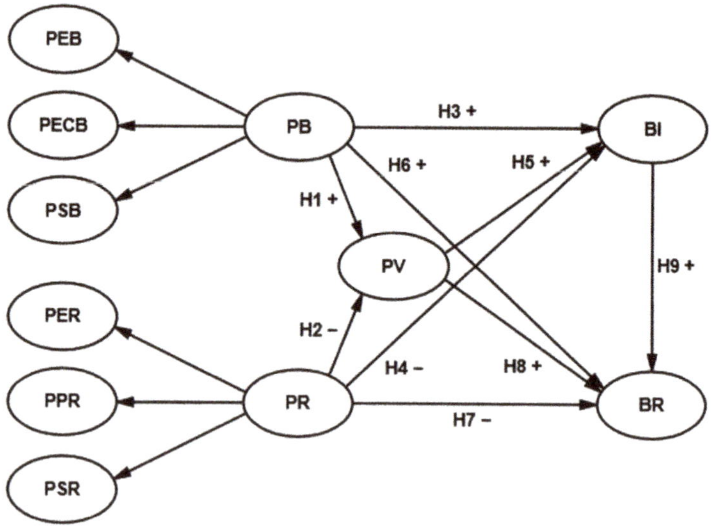

Figure 1. Overall decision-making model of farmers' cultivated land quality protection behavior.

In this study, the perceived benefits, risks, values, and willingness pertaining to farmers' cultivated land quality protection are latent variables that are difficult to predict directly and can be quantified using a structural equation model. At the same time, the structural equation model can deal with the measurement error in the analysis process, thus making the research result increasingly reliable. Therefore, the structural equation model was selected in this study, expressed as follows:

$$Y = \Lambda_Y \eta + \varepsilon \tag{1}$$

$$X = \Lambda_X \xi + \delta \tag{2}$$

$$\eta = \beta \eta + \Gamma \xi + \zeta \tag{3}$$

Equations (1) and (2) are measurement models describing the relationship between latent and observed variables. X is the observed variable of the exogenous latent variable ξ, Y is the observed variable of the endogenous latent variable η, Λ_X and Λ_Y, respectively, represent the factor loading matrix from latent variable ξ and η to the observed variables, and δ and ε, respectively, represent the residual term of exogenous and endogenous variables. Equation (3) is a structural model describing the relationship between latent variables, where η is the endogenous latent variable, ξ is the exogenous latent variable, β denotes the relationship matrix between the endogenous latent variable and the endogenous latent variable, Γ denotes the relationship matrix between the exogenous latent variable and the endogenous latent variable, and ζ is the residual term.

2.3. Variable Selection and Scale Design

In this study, the measurement dimensions and item setting principles of farmers' perceived value of cultivated land quality protection include the following: full reflection of the research question, i.e., the impact of farmers' perceived value on cultivated land quality protection behavior; dimension and item setting to avoid repetition and internal correlation; easily understandable questions; reduction in error due to misunderstanding. On the basis of the analytical and theoretical model of several concepts and constituent dimensions mentioned above, two subscales of willingness and behavior measurement and three subscales of perception measurement are constructed: perceived benefit, perceived risk, and perceived value.

(1) Perceived benefit measurement scale. The multifunctional attribute of cultivated land determines that the protection of cultivated land quality has multiple benefits [60]. Farmers protect cultivated land quality because doing so can bring benefits or values. Theoretically speaking, the production practice of cultivated land protection may produce many results; for example, the improvement of cultivated land fertility increases cultivated land yield and farmers' planting income, the alleviation of soil erosion and pollution improves farmers' production and living environment, and the protection of cultivated land quality boosts the health of agricultural products and national food security.

(2) Perceived risk measurement scale. According to Zeithamld's research, an individual's perceived risk includes perceived monetary risk and perceived nonmonetary risk [44]. Scholars have designed measurement items from two aspects: risk loss of input and expected loss of income. In addition, some scholars measured perceived risk from psychological risk [38,40,43,52]. Therefore, we measured farmers' perceived economic risk of farmland quality protection from three aspects: perceived economic risk, psychological risk, and situational risk. Perceived economic risk mainly refers to the factor cost farmers need to pay for farmland quality protection. Psychological risk mainly refers to the inner anxiety of farmers in the protection of cultivated land quality. Scenario risk mainly refers to farmers' concern about the input uncertainty of farmland quality protection.

(3) Perceived value measurement scale. Farmers' perceived value of cultivated land quality protection refers to the subjective evaluation of farmers' perceived benefits and contributions in cultivated land quality protection decision making [26]. This study draws lessons from the current research results. First, for the protection of farmers' cultivated land quality, it is reflected in the comparison of farmers' perceived benefits and efforts obtained by their behavioral decisions [32]. Second, farmers evaluate whether their cultivated land quality protection behavior meets individual needs [61].

(4) Behavior intentions measurement scale. Farmers' willingness to invest in quality protection of farmland reflects farmers' willingness to protect farmland quality to a certain extent. The level of each factor cost payment that farmers are willing to accept reflects the degree of willingness.

(5) Behavior responses measurement scale. According to the National Agricultural Sustainable Development Plan (2015–2030) and Action Plan for the Protection and Improvement of Cultivated Land Quality, referring to the study of Liu (2018) and Wang (2020) [33,50], the cultivated land quality protection behavior was measured from three aspects: implement conservation farming methods, implement measures to improve soil fertility, and take pollution control and restoration measures.

On the basis of perceived value theory, the variable selection and scale design of existing studies were used for reference, and the results of semi-structured interviews with farmers in the surveyed region were combined. Moreover, the interview results show that farmers generally have a complex of concealing benefits and exaggerating risks. Therefore, in the process of questionnaire design, the variable connotation of perceived benefit was set conservatively, while the variable connotation of perceived risk was relatively radical. In this study, 29 items were designed to measure 11 variables in farmers' land input behavioral decision models using a five-point Likert scale. The specific variable selection and scale design results and their meanings are shown in Table 1.

Table 1. Measurement scale of peasant households' cultivated land quality protection perception.

Variables		Indicators	Questions	Scale Sources	Assignment
Perceived benefits (PB)	Perceived economic benefits (PEB)	Increase in grain output	(PEB1) Quality protection of cultivated land can improve crop yield.	[26,33]	1 = strongly disagree; 2 = don't quite agree; 3 = in general; 4 = agree; 5 = completely agree
		Continued increase in agricultural income	(PEB2) Quality protection of cultivated land can increase agricultural income continuously.		
		Comprehensive cost savings	(PEB3) The comprehensive cost can be reduced by the quality protection of cultivated land.		
	Perceived ecological benefits (PECB)	Improved soil fertility	(PECB1) It is beneficial to improve soil fertility to protect cultivated land quality.		
		Improved ecological environment	(PECB2) Protecting cultivated land quality is beneficial to improving the ecological environment.		
		Reduced water pollution	(PECB3) Quality protection of cultivated land can reduce soil and water pollution.		
	Perceived social benefits (PSB)	Safety and health of agricultural products	(PSB1) Quality protection of cultivated land is beneficial to guarantee the quality and safety of agricultural products.		
		Food security	(PSB2) Quality protection of cultivated land is beneficial to food security.		
		Living security	(PSB3) Quality protection of cultivated land plays an important role in ensuring life in the future.		
Perceived Risks (PR)	Perceived economic risk (PER)	Capital consumption	(PER1) More money is needed to protect the quality of arable land.	[26,62,63]	1 = strongly disagree; 2 = don't quite agree; 3 = in general; 4 = agree; 5 = completely agree
		Labor cost	(PER2) More labor is needed to protect cultivated land quality.		
		Economic benefits	(PER3) The benefits of arable land quality protection appear too slow.		
	Perceived psychological risk (PPR)	Behavior expected	(PPR1) Protection requires participation and the fear that one's own efforts won't work		
		Behavior consequences	(PPR2) Farmland is difficult to improve once it is destroyed.		
		Knowledge of technology	(PPR3) I worry that I can't master relevant knowledge and technology.		
	Perceived situational risk (PSR)	Policy scenarios	(PSR1) I am concerned that relevant policies are not in place.		
		Uncertainty	(PSR2) I am afraid to try for fear of failure.		
		Disaster risk	(PSR3) Fear of losses due to sudden weather disasters and pests.		

Table 1. *Cont.*

Variables	Indicators	Questions	Scale Sources	Assignment
Perceived value (PV)		(PV1) Protecting cultivated land quality is of great significance. (PV2) I hold a positive attitude toward the protection of cultivated land quality. (PV3) Carrying out cultivated land quality protection brings me certain benefits.	[26,64]	1 = strongly disagree; 2 = don't quite agree; 3 = in general; 4 = agree; 5 = completely agree
Behavior intentions (BI)		(BI1) I am willing to put in work to protect the quality of cultivated land. (BI2) I am willing to invest energy and time to protect the quality of cultivated land. (BI3) I am willing to invest money to protect the quality of cultivated land.	[26,50]	1 = strongly disagree; 2 = don't quite agree; 3 = in general; 4 = agree; 5 = completely agree
Behavior responses (BR)		(BR1) Implement conservation farming methods. (BR2) Implement measures to improve soil fertility. (BR3) Take pollution control and restoration measures.	[25,50]	1 = never; 2 = rarely; 3 = sometimes; 4 = often; 5 = always

3. Variable Measurement and SEM Model Data Verification

3.1. Data Sources

Ningbo City is located at 120°55′ to 122°16′ E longitude and 28°51′ to 30°33′ N latitude. It is located in the middle of China's coastline in the southern wing of the Yangtze River Delta, and it is responsible for the jurisdiction of Haishu, Jiangbei, Zhenhai, Beilun, Yinzhou, Fenghua (six districts), Ninghai, Xiangshan (two counties), Cixi, and Yuyao (two county-level cities). The terrain is high in the southwest and low in the northeast. The city's geomorphology is divided into mountains, hills, platforms, valleys (basins), and plains, of which the plains account for 40.3%. The region has a subtropical monsoon climate, mild and humid, with four distinct seasons. The city's annual average temperature is 16.4 °C, the average yearly precipitation is about 1480 mm, and the average annual sunshine duration is 1850 h, which is suitable for the growth of crops. Ningbo is the main distribution area of cultivated land resources in Zhejiang Province. It is an essential commercial grain production base and the main supply base of crops, such as grain, oil, vegetables, tea, and fruit. Hence, it has a good representation in the selection of research areas.

The data used in this study come from the survey of farmer households conducted by the research team in Ningbo from May to August 2021. A stratified proportional random sampling method was adopted for experimental investigation sampling. Given the comprehensiveness of the data and the authenticity of the reflections, three ecological civilization areas and two common areas were randomly selected from six ecological civilization demonstration areas and four common areas. The ecological civilization demonstration areas were Zhenhai District, Beilun District, and Xiangshan County, while the ordinary districts were Yinzhou District and Yuyao District, indicating a total of five counties (districts). Then, three administrative villages were selected from each county (district) with 15 villages, and 20 farmers were randomly selected from each village as survey samples. To improve the accuracy of data collection, data collection was conducted by investigators trained by the research group in a one-to-half structured interview. After eliminating the samples with contradictory and incomplete information, 288 effective samples were obtained, with an efficiency rate of 96%. The characteristics of the sample farmers are shown in Table 2.

Table 2. Characteristics of sample farmers.

Variable	Classification Criteria	Frequency (Times)	Frequency (%)	Variable	Classification Criteria	Frequency (Times)	Frequency (%)
Sex	Male	218	75.69	Age	Under 35	2	0.7
	Female	70	24.31		35–45	19	6.62
Identity	Ordinary villager	217	75.35		45–55	63	21.95
	Village cadres	40	13.89		55–65	100	34.84
	Party member	31	10.76		Over 65	103	35.89
Education	Primary and below	144	50	Scale of contracted land	2 acres and below	116	40.28
	Junior high school	85	29.51		2–5 acres	87	30.21
	Vocational high school or high school	42	14.58		5–10 acres	24	8.33
	College	15	5.21		10–20 acres	20	6.94
	Bachelor's degree and above	2	0.69		Over 20 acres	41	14.24
Type of employment	Agriculture	160	55.56	Share of agricultural labor force	0–25%	156	54.2
	Agriculture-oriented and part-time	47	16.32		26–50%	63	21.9
	Job-oriented and part-time	75	26.04		51–75%	26	9
	Retired	6	2.08		76–100%	43	14.9

3.2. Cultivated Land Quality Protection Perceived Value Scale and Its Statistical Description

On the basis of the theory of perceived value, drawing on the variable selection and scale design results of existing research, and combining the results of semi structured interviews with farmers in the survey area, we adopted a 5-point Likert scale, where 1 = "completely disagree", 2 = "disagree", 3 = "general", 4 = "agree", and 5 = "completely agree", to measure the variables in the decision-making model of the cultivated land quality protection behavior of farmers. The cultivated land quality protection behavior was measured from three aspects, namely, "I implement conservation farming methods", "I implement measures to improve soil fertility", and "I take pollution control and restoration measures". According to farmers' responses to the frequency of relevant measures, the same five-level assignment of "1–5" denoting "never", "rarely", "sometimes", "often", and "always", respectively, was adopted. The specific variable selection, scale design results, and their descriptive statistical results are shown in Table 3.

Farmers generally have a relatively high-interest perception of cultivated land quality protection. The average scores of economic interest, ecological interest, and social interest perception were all at a high level: 3.74, 3.79, and 3.86, respectively. The scores of the perceived social and ecological benefits were higher than those of the perceived economic benefits. Specifically, farmers' perception of the benefits of cultivated land quality protection behavior mainly focuses on its essential role in ensuring future life, ensuring the quality and safety of agricultural products, improving the ecological environment, and improving crop yield. This shows that farmers subconsciously recognize cultivated land quality protection behavior and think that its value lies in economic benefits. They also consider the positive externalities of cultivated land quality protection behavior from social and ecological aspects.

Table 3. Variable table and its descriptive statistics.

Variables	Items	Mean	Standard Deviation
Perceived economic benefits (PEB)	Carrying out cultivated land quality protection can increase crop yields.	3.889	0.938
	Carrying out cultivated land quality protection can continuously increase agriculture income.	3.667	1.029
	Carrying out cultivated land quality protection can save comprehensive costs.	3.667	0.933
Perceived ecological benefits (PECB)	Carrying out cultivated land quality protection is conducive to improving soil fertility.	3.771	1.000
	Protecting the quality of cultivated land is conducive to improving the ecological environment.	3.806	1.007
	Carrying out cultivated land quality protection can reduce water and soil pollution.	3.788	1.026
Perceived society benefits (PSB)	Protecting the quality of cultivated land is conducive to ensuring the quality and safety of agricultural products.	3.899	1.002
	Protecting the quality of cultivated land is conducive to national food security.	3.729	1.051
	Carrying out the protection of cultivated land quality has an essential role in ensuring future life.	3.941	1.002
Perceived economic risks (PER)	More money is needed to protect the quality of cultivated land.	3.708	1.110
	More labor is needed to protect the quality of cultivated land.	3.694	1.119
	The benefits of cultivated land quality protection appear too slow.	3.677	1.152
s	Implementation of protection requires joint participation in the fear that self-effort will not be effective.	3.194	1.017
	Worried that once the cultivated land is destroyed, it will be difficult to improve.	3.656	1.188
	Worried about not being able to master relevant knowledge and technology.	3.337	1.151
Perceived scenario risk (PSR)	I am worried about the inadequate implementation of relevant policies.	3.941	0.998
	I am afraid to try fear of failure.	3.913	1.044
	I am worried about losses due to sudden weather disasters and pests.	3.663	1.016
Perceived value (PV)	Protecting the quality of cultivated land is of great significance.	3.483	0.805
	I am optimistic about protecting the quality of cultivated land.	3.424	0.840
	Carrying out cultivated land quality protection has brought me certain benefits.	3.302	0.897
Behavior intentions (BI)	I am willing to put in work to protect the quality of cultivated land.	3.462	0.906
	I am willing to invest energy and time to protect the quality of cultivated land.	3.378	0.994
	I am willing to invest money to protect the quality of cultivated land.	3.083	0.977
Behavior responses (BR)	I implement conservation farming methods.	2.646	1.052
	I implement measures to improve soil fertility.	2.642	1.133
	I take pollution control and restoration measures.	2.670	1.001

In terms of risk perception, farmers' risk perception of cultivated land quality protection behavior was relatively high. The average economic risk, psychological risk, and situational risk perception scores were 3.69, 3.4, and 3.84, respectively. The scores of situation risk and economic risk were higher than those of psychological risk. Specifically, farmers' risk perception of cultivated land quality protection behavior mainly comes from their concerns about the prospect and sustainability of relevant policies and the failure of decision making on cultivated land quality protection behavior to cope with the uncertainty of the future, which is a high error cost. The scores of these two perceived scenario risk options were 3.94 and 3.91, respectively. This situation also shows that at this stage, situational risk is a factor that cannot be ignored. In the face of future and policy uncertainties, farmers, as rational people, will preferentially choose to avoid risks.

The scores of perceived value, intention, and behavior showed a trend of gradual decline, with the average scores being 3.4, 3.31, and 2.65, respectively. Although farmers recognize and perceive the value of cultivated land quality protection behavior, a sign of "fading enthusiasm" is reflected in the willingness to engage in cultivated land quality protection. In choosing specific protection behavior, farmers show a prominent characteristic of "do not mind". In short, farmers subconsciously recognize cultivated land quality protection behavior and its significance. However, they have many concerns related to internal and external factors in the actual cultivated land quality protection behavior. In this study, the probability of choosing cultivated land quality protection behavior was low, and the overall score was >3.

3.3. Scale Reliability and Validity Test

To measure the reliability and correctness of the data, we tested the reliability and validity of the survey data. The results are shown in Table 4. In the reliability test, Cronbach's coefficient value of each latent variable was between 0.804 and 0.892, thus meeting the required value greater than 0.7. The combined reliability (CR) was more significant than 0.7, indicating that the scale had good internal consistency. The KMO value of each latent variable in the validity test was between 0.677 and 0.736, thus meeting the required value greater than 0.5. The significance level of the chi-square value in the Bartlett sphere test was significant. The load coefficients on the common factors of all observed variables ranged from 0.720 to 0.915, thus meeting the required value greater than 0.6. The above results show that the model data had good reliability and validity and that the data quality passed the test.

Table 4. Reliability and validity test results.

Variables	Items	Standard Factor Load	Reliability of Questions	Cronbach Coefficient	Component Reliability		Convergent Validity
		Std.	SMC	Cronbach's α	CR	KMO	AVE
PEB	PEB1	0.816	0.666		0.806	0.709	0.582
	PEB2	0.741	0.549	0.804			
	PEB3	0.729	0.531				
PECB	PECB1	0.720	0.518		0.849	0.701	0.654
	PECB2	0.915	0.837	0.844			
	PECB3	0.780	0.608				
PSB	PSB1	0.761	0.579		0.818	0.714	0.600
	PSB2	0.765	0.585	0.817			
	PSB3	0.798	0.637				
PER	PER1	0.864	0.746	0.873	0.874	0.736	0.698
	PER2	0.790	0.624				
	PER3	0.850	0.723				
PPR	PPR1	0.789	0.623	0.819	0.824	0.715	0.610
	PPR2	0.733	0.537				
	PPR3	0.818	0.669				

Table 4. *Cont.*

Variables	Items	Standard Factor Load	Reliability of Questions	Cronbach Coefficient	Component Reliability		Convergent Validity
		Std.	SMC	Cronbach's α	CR	KMO	AVE
PSR	PSR1	0.772	0.596	0.864	0.866	0.729	0.683
	PSR2	0.860	0.740				
	PSR3	0.844	0.712				
PV	PV1	0.742	0.551	0.818	0.828	0.677	0.617
	PV2	0.862	0.743				
	PV3	0.747	0.558				
BI	BI1	0.856	0.733	0.892	0.896	0.729	0.742
	BI2	0.890	0.792				
	BI3	0.837	0.701				
BR	BR1	0.795	0.632	0.852	0.854	0.732	0.661
	BR2	0.836	0.699				
	BR3	0.807	0.651				

3.4. Model Fit Test

AMOS 23.0 and the maximum likelihood method were used to estimate the model's parameters. The fitting indices of the decision-making model of farmland quality protection, including the absolute-fit index, value-added fit index, and reduced-fit index, were calculated. The results are shown in Table 5. All model indicators reached the normal state, and the overall fitting effect was good.

Table 5. Results of overall model fit test.

Categories	Indicators	Adapter Standard	Statistics	Adaptation to Judge
Absolute fit index	χ^2/DF	<3	1.120	Good
	GFI	>0.9	0.919	Good
	AGFI	>0.9	0.901	Good
	RMSEA	<0.5	0.020	Good
	RMR	<0.5	0.035	Good
Value-added compatibility indicators	TLI	>0.9	0.991	Good
	CFI	>0.9	0.992	Good
	IFI	>0.9	0.992	Good
	NFI	>0.9	0.932	Good
	RFI	>0.9	0.922	Good
Simple fit index	PGFI	>0.5	0.749	Good
	PNFI	>0.5	0.817	Good

4. Results and Hypothesis Verification

4.1. Path Result Estimation for Structural Models

The structural equation model results confirmed the Hypotheses H1–H9, thus indicating that farmers' behavior decision for farmland quality protection conforms to the perceived value theory. Farmers' cultivated land quality protection behavior followed the decision logic path of "perceived value→behavioral intention→behavioral expression". Farmers' cultivated land quality protection behavior responses (BR) were influenced by willingness (BI), perceived value (PV), perceived benefit (PB), perceived risk (PR), and other latent variables. Among them, perceived benefit (PB), perceived risk (PV), and perceived value (PV) all directly affected farmers' behavior (BR). At the same time, willingness (BI) played an intermediary role among perceived value (PV), perceived benefit (PB), perceived risk (PV), and behavioral response (BR). Perceived value (PV) played an intermediary role between perceived benefit (PB) and perceived risk (PV).

As shown in Figure 2 and Table 6, perceived benefit (PB) and perceived risk (PR) were the main factors affecting perceived value (PV), and their standardized path coefficients were 0.528 and −0.409, respectively, significant at the statistical level of 1%. The results show that the perceived benefits of farmland quality protection had a significant positive impact on farmers' perceived value. The perceived risk has a significant negative impact on farmers' perceived value. Hypotheses H1 and H2 were thus assumed to be verified.

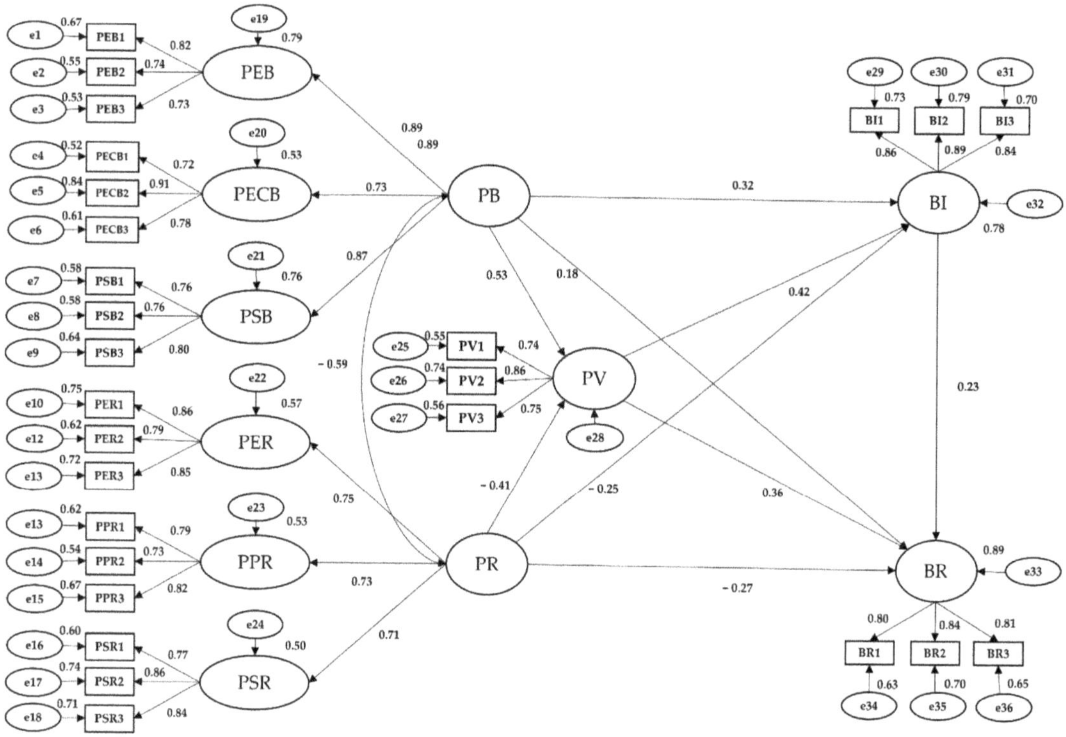

Figure 2. Structural equation model path and coefficient.

Table 6. Model path coefficient and hypothesis testing.

Influence Path	Estimate	S.E.	C.R	P	Hypothesis Testing
Perceived benefit→perceived value	0.528	0.073	6.357	***	Accepted H1
Perceived risk→perceived value	−0.409	0.070	−4.847	***	Accepted H2
Perceiving interest→willingness	0.361	0.096	3.771	***	Accepted H3
Perceived risk→willingness	−0.249	0.087	−3.086	**	Accepted H4
Perceived value→willingness	0.416	0.136	3.988	***	Accepted H5
Perceived benefit→behavior	0.177	0.101	2.157	*	Accepted H6
Perceived risk→behavior	−0.273	0.095	−3.336	***	Accepted H7
Perceived value→behavior	0.363	0.147	3.448	***	Accepted H8
Willingness→behavior	0.232	0.112	2.232	*	Accepted H9

Note: Significant at the * 10%, ** 5%, and *** 1% levels.

Farmers' perceived benefit (PB), perceived risk (PR), and perceived value (PV) were the main factors affecting farmers' willingness (BI); the standardized path coefficients were 0.361, −0.249, and 0.416, respectively, significant at the statistical level of 1%. The results show that farmland quality protection's perceived benefits and values significantly influence farmers' willingness. By contrast, the perceived risks had a significant negative impact on farmers' willingness. Thus, Hypotheses H3, H4, and H5 were verified.

Farmers' perceived benefit (PB), perceived risk (PR), perceived value (PV), and willingness (BI) were the main factors affecting farmers' behavior (BR). The standardized path coefficients were 0.177, −0.273, 0.363, and 0.232, significant at the statistical levels of 5%, 1%, 1%, and 5%, respectively. Hence, Hypotheses H6–H9 were verified.

In sum, the causal relationship between potential variables was significant. The behavioral decision model constructed in this study could effectively explain the internal mechanism of farmers' cultivated land quality protection behavior.

4.2. Mediation Effect Analysis

The potential variables' direct, indirect, and total effects were summarized to indicate the interaction between potential variables. The results after standardized treatment are shown in Table 7.

Table 7. Direct effect, indirect effect, and total effect among latent variables.

Latent Variables	Direct Effect	Indirect Effect	Total Effect
Perceived benefit→perceived value	0.528	-	0.528
Perceived risk→perceived value	−0.409	-	−0.409
Perceiving interest→willingness	0.316	0.220	0.536
Perceived risk→willingness	−0.249	−0.170	−0.420
Perceived value→willingness	0.416	-	0.416
Perceived benefit→behavior	0.177	0.316	0.493
Perceived risk→behavior	−0.273	−0.246	−0.519
Perceived value→behavior	0.363	0.096	0.459
Willingness→behavior	0.232	-	0.232

Regarding the perceived value of farmland quality protection, perceived benefit significantly affected perceived value. By contrast, perceived risk had a significant negative effect on perceived value. Farmers' perceived value of cultivated land quality protection resulted from a two-way tradeoff between perceived benefit and risk. The impact of farmers' perceived benefit (0.528) on their perceived value was greater than that of perceived risk (−0.409). The standardized path coefficients of perceived benefits and the next three endogenous variables, namely, farmers' perception of economic, ecological, and social benefits, were 0.89, 0.73, and 0.87, respectively. The result shows that farmers' stronger perceived benefits would improve their perceived value. The standardized path coefficients of perceived risk and its next three endogenous variables reflecting farmers' perception of economic, psychological, and psychological risks were 0.75, 0.73, and 0.71, respectively. The result indicates that farmers still had some worries in the decision-making process for cultivated land quality protection. When farmers protect farmland, uncertain factors such as capital cost, inability to master relevant technologies, and fear of losses would have the greatest impact on them, thus reducing farmers' value perception.

From the relative magnitude of each factor on farmers' willingness to protect cultivated land quality, the direct effects of perceived benefits and perceived risks were 0.316 and −0.249, and the indirect effects were 0.220 and −0.170, respectively. Both had important direct and indirect effects on farmers' willingness to protect cultivated land quality. The absolute value of perceived benefits was greater than the absolute value of perceived risks. This result shows that the impact of perceived benefits on farmers' willingness to protect cultivated land quality was greater than that of perceived benefits. The same was true for the impact on perceived value. This proves that farmers are sensitive to the benefits of cultivated land quality protection in the early stage of behavioral decision making. Before making the behavioral decision for cultivated land quality protection, farmers preliminarily assess the possible benefits and their expectations. Greater perceived economic benefits of cultivated land lead to a greater willingness to protect cultivated land quality. This result is in line with the hypothesis of "rational economic individuals", that is, as rational economic individuals, farmers aim to maximize interests. In addition, the direct effect of perceived value on farmland quality protection was the largest (0.416). This indicates that farmers' strong perceived value would enhance the willingness to protect cultivated land quality.

From the behavioral effects of each factor on the actual farmland quality protection of farmers, the direct effects of perceived benefits and perceived risks were 0.177 and −0.273, and the total effects were 0.493 and −0.519, respectively. Farmers' perceived benefits and perceived risks exert significant effects on their cultivated land quality protection behaviors, and the absolute value of perceived risks was greater than that of perceived benefits. This result indicates that perceived risk had a greater impact on farmers' cultivated land quality behavior than perceived benefit. Compared with the benefits of cultivated land quality protection, the possible risks and losses in the actual behavior stage pose a greater concern for farmers. Farmers generally have "risk aversion" psychology when faced with the choice of whether to protect cultivated land quality, and they are typical "risk avoiders" in the behavior of protecting cultivated land quality. This result also explains why the value of farmers' cultivated land quality protection behavior was lower than their value and willingness in previous variable statistics. Meanwhile, the direct effect of perceived value was 0.363, which was the largest among all factors. The results show that perceived value was the most important direct factor affecting farmers' cultivated land quality protection behavior.

5. Discussion

Although farmers' cultivated land quality protection behavior is affected by many factors, one should clarify the formation of farmers' perceived value and recognize the role of perceived factors in it. Studies have shown that farmers' perceived value of cultivated land quality protection results from a comprehensive tradeoff between perceived benefits and perceived risks. Wang and Guo (2020) pointed out that when farmers participate in cultivated land quality protection, the first consideration is whether they can obtain considerable benefits [33].

For farmers, whether they can improve crop yield, continuously increase agricultural income, and save total cost are the primary factors they consider when making decisions on farmland quality protection. The guarantee of future life and the quality and safety of agricultural products are also critical. At the same time, farmers have begun to pay attention to improving soil fertility, ecological environment, and other ecological benefits. Although the reasonable goal of realizing the maximization of economic benefits has not changed, the rational behavior structure of farmers may have begun to change. Economic benefits are no longer the most appealing to farmers; social and ecological benefits are also worthy of attention [59,60].

However, in terms of the quality of the cultivated land protection decisions of farmers, such as the use of organic fertilizers, soil fertilizers, and soil ameliorant, farmers need to pay the cost first. With income uncertainty, farmers, as a "rational economic individuals", avoid actions that may cause maximum welfare loss. In other words. They minimize the maximum welfare loss while making decisions, which tend to follow the "minimum, maximum principle" in the face of risk and uncertainty [36]. Therefore, improving farmers' perceived benefits, especially monetary income benefits, and reducing perceived risks, especially cost input risks, can help improve their perceived value level of cultivated land quality protection [38]. In addition, the direct effect of perceived value on farmers' willingness and behavior of farmland quality protection is far greater than that of perceived benefit and perceived risk. Perceived value is the most critical direct factor affecting farmers' cultivated land quality protection behavior, as confirmed by some research in the field of farmers' behavior [26]. Therefore, improving farmers' perceived value level is helpful to stimulate their willingness to protect cultivated land quality in practice and improve their cultivated land quality protection behavior.

In this study, we integrated farmers' willingness and behavior of farmland protection into a unified analytical framework. The results show that farmers' willingness to engage in farmland quality protection was generally positive, but their actual behavior was significantly lower than their willingness. Some studies on green agricultural production and ecological farming have clarified this point of view [53]. At the same time, the impact of perceived benefits and perceived risks on farmers' willingness and behavior is asymmetrical [26]. According to the calculation results of latent direct effect, indirect effect and total effect, compared with perceived risk, perceived benefit had more influence on the formation of farmers' perceived value (direct effect of perceived benefit 0.528 vs. direct effect of perceived risk −0.409), and perceived benefit has the greatest influence on farmers' willingness to protect cultivated land quality. However, perceived risk had the greatest influence on farmers' cultivated land quality protection behavior. The direct effect of farmers' perceived risk on farmers' behavior was −0.273, higher than that of farmers' perceived benefit (0.177). In the initial stage of farmland quality protection, farmers are highly sensitive to their benefits. Nevertheless, the occurrence of actual behavior is affected by the perceived risk. The possible explanation is that agriculture itself is a high-risk industry [65]. The instability of macro- and microeconomic environments and natural conditions makes agricultural production face many risks and uncertainties. However, in terms of the protection of cultivated land quality, some factors will bring more uncertainty and higher risks to farmers. Examples include whether deep tillage, soil improvement, crop rotation, and other protective tillage methods are adopted; the implementation of organic fertilizer, green fertilizer, or under the soil test formula, and other measures to cultivate fertilizer; and pollution control and remediation measures, such as reducing the amount of fertilizer and applying low-toxicity and low-residue pesticides. Although the first consideration of farmers is to obtain considerable income before taking action, the improvement of family life brought by higher-than-expected income may not be enough to compensate for the devastating impact brought by lower-than-expected income. For farmers, the utility of high income is far below the stability of utility. This view relates to that of Scott (1976), who reported that most farmers in the developing world follow the principle of "safety first" for the production of life [51]. Under the survival ethic of "safety first", farmers do not pursue income maximization. A vast majority of farmers can be said to belong to the risk-aversion type in the protection behavior of cultivated land quality.

The unexpected finding is that having agricultural insurance can significantly reduce farmers' risk perception. In the survey sample, the risk perception of farmers with agricultural insurance is much lower than those without insurance. Their behavior value is also higher, which provides a new idea for us to formulate relevant policies. If only farmers adopted targeted policies or measures, such as subsidies, the regulation of cultivated land quality protection may not be able to achieve the desired effect. To help farmers disperse or transfer risks, they should be encouraged to buy agricultural insurance actively while promoting agricultural insurance premiums and improving their risk defense [38,40,66]. Such support will benefit the quality of farmers' cultivated land protection behavior decisions.

The research offers the following policy enlightenment. First, policymakers and experts should give full play to grassroots initiatives, combine modern science and technology means, increase publicity, and improve farmers' awareness of farmland quality protection so that farmers will have a profound understanding of the ecological and social value of farmland and improve their perceived value level. Second, the intensity of incentives should be increased further by strengthening agricultural subsidies or other preferential policies, as well as the economic interests of farmers. Experts should also improve their awareness and the pricing mechanism for agricultural products and materials. Moreover, they should safeguard farmers' income to ensure latter's motivation toward arable land protection. Third, the coverage and compensation intensity of various kinds of insurance, especially agricultural insurance, in rural areas should be improved. Loss aversion is an important reason for farmers' risk avoidance. Hence, improving insurance coverage can help reduce farmers' losses and improve their ability to cope with risks.

6. Conclusions

To improve the cultivated land quality protection system and improve the system's performance, this study investigates farmers' cultivated land quality protection behavior from the aspect perception. The internal logical mechanism of farmers' cultivated land quality protection behavior is explored. From the perspective of farmers' perception and on the basis of the theory of perceived value and behavior of farmers related to the quality of the sample cultivated land protection area, this study builds the "overall farmers' perceived quality of cultivated land protection behavior decision model" through the internal factors of the model parameter estimation. An analysis of peasant household perception factors and their influence on behavior decisions explains the internal logic of farmers' protection behavior decision-making mechanism in protecting the quality of cultivated land.

On the basis of the analysis, the following conclusions can be drawn. First, the quality of farmers' arable land protection behavior is based on the perceived value of the result of rational decision making. Its action logic follows "perceived balance to perceived value, behavior intention, and behavior response", which is the path to the paradigm of farmers' perception of the quality of cultivated land protection value to protect behavior and produce the critical nature of direct and indirect influence. Second, farmers' perceived value of cultivated land quality protection results from a total balance between their perceived benefits and risks. Improving farmers' perceived benefits, especially monetary income benefits, and reducing perceived risks and significant cost input risks will help improve farmers' perceived value of cultivated land quality protection. Third, farmers' perceived benefits and risks significantly impact their willingness and behavior of farmland quality protection. Farmers are sensitive to benefits in the initial stage of farmland quality protection, but their actual behavior is greatly affected by perceived risks, and they are typical "risk avoiders". The results indicate that most of the surveyed farmers can recognize the basic value of cultivated land quality protection. Farmers have potential enthusiasm for the protection of cultivated land quality. However, in actual decision making, they are often more sensitive to risk factors. Among them, factor cost, inner anxiety and uncertainty are the main inducements hindering their behavior.

Author Contributions: Conceptualization, X.W. (Xiaoying Wang) and A.N.; Investigation, X.W. (Xiaoying Wang) and H.H.; Methodology, H.H.; Writing—original draft, X.W. (Xiaoying Wang) and A.N.; Writing—review & editing, A.N., G.L. and X.W (Xueqi Wang). All authors have read and agreed to the published version of the manuscript.

Funding: This research was funded by Zhejiang Provincial Natural Science Foundation of China, grant number LQ21G030003, LQ22G030001 and by the Fundamental Research Funds for the Provincial Universities of Zhejiang, Grant number SJWY2021002, SJWY2022004, and by Ningbo Social Science Research Base Project, Grant number ZX2019001133, JD5-ZD41, and by College Students Science and Technology Innovation Activity Plan (New Seedling Talent Plan) Project of Zhejiang Province, Grant number 2021R405090.

Informed Consent Statement: Informed consent was obtained from all subjects involved in the study.

Data Availability Statement: The data presented in this study are available on request from the corresponding author. The data are not publicly available due to privacy.

Conflicts of Interest: The authors declare no conflict of interest.

References

1. Viana, C.M.; Freire, D.; Abrantes, P.; Rocha, J.; Pereira, P. Agricultural land systems importance for supporting food security and sustainable development goals: A systematic review. *Sci. Total Environ.* **2022**, *806*, 150718. [CrossRef]
2. Zhou, Y.; Li, X.; Liu, Y. Cultivated land protection and rational use in China. *Land Use Policy* **2021**, *106*, 105454. [CrossRef]
3. Liu, Y.; Zhou, Y. Reflections on China's food security and land use policy under rapid urbanization. *Land Use Policy* **2021**, *109*, 105699. [CrossRef]
4. Chen, L.; Zhao, H.; Song, G.; Liu, Y. Optimization of cultivated land pattern for achieving cultivated land system security: A case study in Heilongjiang Province, China. *Land Use Policy* **2021**, *108*, 105589. [CrossRef]
5. Huang, L.; Feng, Y.; Zhang, B.; Hu, W. Spatio-Temporal Characteristics and Obstacle Factors of Cultivated Land Resources Security. *Sustainability* **2021**, *13*, 8498. [CrossRef]
6. Zhang, Z.; Meng, X.; Elahi, E. Protection of Cultivated Land Resources and Grain Supply Security in Main Grain-Producing Areas of China. *Sustainability* **2022**, *14*, 2808. [CrossRef]

7. Zhao, Q.; Zhou, S.; Wu, S.; Ren, k. Cultivated Land Resources And Strategies for its Sustainable Utilization And Protection in China. *Acta Pedol. Sin.* **2006**, *43*, 662–672, (In Chinese with English abstract).
8. Wu, Y.; Shan, L.; Guo, Z.; Peng, Y. Cultivated land protection policies in China facing 2030: Dynamic balance system versus basic farmland zoning. *Habitat Int.* **2017**, *69*, 126–138. [CrossRef]
9. Liang, C.; Penghui, J.; Wei, C.; Manchun, L.; Liyan, W.; Yuan, G.; Yuzhe, P.; Nan, X.; Yuewei, D.; Qiuhao, H. Farmland protection policies and rapid urbanization in China: A case study for Changzhou City. *Land Use Policy* **2015**, *48*, 552–566. [CrossRef]
10. Lichtenberg, E.; Ding, C. Assessing farmland protection policy in China. *Land Use Policy* **2008**, *25*, 59–68. [CrossRef]
11. Jikun, H.; Lifen, Z.; Xiangzheng, D.; Scott, R. Cultivated land changes in China: The impacts of urbanization and industrialization. In Proceedings of the Proc. SPIE, San Diego, CA, USA, 1 September 2005.
12. Ge, D.; Long, H.; Zhang, Y.; Ma, L.; Li, T. Farmland transition and its influences on grain production in China. *Land Use Policy* **2018**, *70*, 94–105. [CrossRef]
13. Van Vliet, J.; Eitelberg, D.A.; Verburg, P.H. A global analysis of land take in cropland areas and production displacement from urbanization. *Glob. Environ. Chang.* **2017**, *43*, 107–115. [CrossRef]
14. Reenberg, A. Agricultural land use pattern dynamics in the Sudan–Sahel—towards an event-driven framework. *Land Use Policy* **2001**, *18*, 309–319. [CrossRef]
15. King, R.; Killingbeck, J. Agricultural and land use change in central Basilicata: From Carlo Levi to the Comunitá Montana. *Land Use Policy* **1990**, *7*, 7–26. [CrossRef]
16. Zhang, H.; Zhang, Y.; Wu, S.; Cai, R. The Effect of Labor Migration on Farmers' Cultivated Land Quality Protection. *Sustainability* **2020**, *12*, 2953. [CrossRef]
17. Wu, S.; Ye, Y.; Lin, Y. Experience and Enlightenment of Multifunctional Land Consolidation in Germany, Japan and Taiwan in China. *J. Huazhong Agric. Univ. (Soc. Sci. Ed.)* **2019**, *3*, 140–148+165–166. [CrossRef]
18. Goodenough, R. Room to grow? Farmland conservation in California. *Land Use Policy* **1992**, *9*, 21–35. [CrossRef]
19. Zhong, T.; Huang, X.; Zhang, X.; Scott, S.; Wang, K. The effects of basic arable land protection planning in Fuyang County, Zhejiang Province, China. *Appl. Geogr.* **2012**, *35*, 422–438. [CrossRef]
20. Ward, R.M. The US Farmland Protection Policy Act: Another case of benign neglect. *Land Use Policy* **1991**, *8*, 63–68. [CrossRef]
21. Wang, B.; Li, F.; Feng, S.; Shen, T. Transfer of development rights, farmland preservation, and economic growth: A case study of Chongqing's land quotas trading program. *Land Use Policy* **2020**, *95*, 104611. [CrossRef]
22. Zhang, W.; Wang, W.; Li, X.; Ye, F. Economic development and farmland protection: An assessment of rewarded land conversion quotas trading in Zhejiang, China. *Land Use Policy* **2014**, *38*, 467–476. [CrossRef]
23. Xie, H.; Huang, Y. Influencing factors of farmers' adoption of pro-environmental agricultural technologies in China: Meta-analysis. *Land Use Policy* **2021**, *109*, 105622. [CrossRef]
24. Chen, J.; Yu, Z.; Ouyang, J.; van Mensvoort, M.E.F. Factors affecting soil quality changes in the North China Plain: A case study of Quzhou County. *Agric. Syst.* **2006**, *91*, 171–188. [CrossRef]
25. Bindraban, P.S.; Stoorvogel, J.J.; Jansen, D.M.; Vlaming, J.; Groot, J.J.R. Land quality indicators for sustainable land management: Proposed method for yield gap and soil nutrient balance. *Agric. Ecosyst. Environ.* **2000**, *81*, 103–112. [CrossRef]
26. Li, M.; Wang, J.; Zhao, P.; Chen, K.; Wu, L. Factors affecting the willingness of agricultural green production from the perspective of farmers' perceptions. *Sci. Total Environ.* **2020**, *738*, 140289. [CrossRef]
27. Chen, S.; Huang, Z. Will collective ownership hinder the protection of cultivated land quality?—Study on farmers' cultivated land quality protection behavior based on cognitive perspective. *Qinghai Soc. Sci.* **2013**, *2*, 7–14, (In Chinese with English abstract). [CrossRef]
28. Liu, H.; Zhou, Y. Urbanization, Land Use Behavior and Land Quality in Rural China: An Analysis Based on Pressure-Response-Impact Framework and SEM Approach. *Int. J. Environ. Res. Public Health* **2018**, *15*, 2621. [CrossRef]
29. Yu, Z.; Yao, L.; Wu, M. Farmers' attitude towards the policy of remediation during fallow in soil fertility declining and heavy metal polluted area of China. *Land Use Policy* **2020**, *97*, 104741. [CrossRef]
30. Akter, M.; Fan, L.; Rahman, M.M.; Geissen, V.; Ritsema, C.J. Vegetable farmers' behaviour and knowledge related to pesticide use and related health problems: A case study from Bangladesh. *J. Clean. Prod.* **2018**, *200*, 122–133. [CrossRef]
31. Hettig, E.; Lay, J.; Sipangule, K. Drivers of Households' Land-Use Decisions: A Critical Review of Micro-Level Studies in Tropical Regions. *Land* **2016**, *5*, 32. [CrossRef]
32. Sapbamrer, R.; Thammachai, A. A Systematic Review of Factors Influencing Farmers' Adoption of Organic Farming. *Sustainability* **2021**, *13*, 3842. [CrossRef]
33. Wang, Q.; Guo, X. Perceived Benefits, Social Network and Farmers' Behavior of Cultivated Land Quality Protection: Based on 410 Grain Growers' Surveys in Hua County, Henan Province. *China Land Sci.* **2020**, *34*, 43–51, (In Chinese with English abstract). [CrossRef]
34. Faridi, A.A.; Kavoosi-Kalashami, M.; Bilali, H.E. Attitude components affecting adoption of soil and water conservation measures by paddy farmers in Rasht County, Northern Iran. *Land Use Policy* **2020**, *99*, 104885. [CrossRef]
35. Xu, M.; Zhang, Z. Farmers' knowledge, attitude, and practice of rural industrial land changes and their influencing factors: Evidences from the Beijing-Tianjin-Hebei region, China. *J. Rural. Stud.* **2021**, *86*, 440–451. [CrossRef]
36. Morais, M.; Borges, J.A.R.; Binotto, E. Using the reasoned action approach to understand Brazilian successors' intention to take over the farm. *Land Use Policy* **2018**, *71*, 445–452. [CrossRef]

37. Engel, S.; Pagiola, S.; Wunder, S. Designing payments for environmental services in theory and practice: An overview of the issues. *Ecol. Econ.* **2008**, *65*, 663–674. [CrossRef]
38. Gong, Y.; Baylis, K.; Kozak, R.; Bull, G. Farmers' risk preferences and pesticide use decisions: Evidence from field experiments in China. *Agric. Econ.* **2016**, *47*, 411–421. [CrossRef]
39. Saqib, S.E.; Ahmad, M.M.; Panezai, S. Landholding size and farmers' access to credit and its utilisation in Pakistan. *Dev. Pract.* **2016**, *26*, 1060–1071. [CrossRef]
40. Pan, D.; He, M.; Kong, F. Risk attitude, risk perception, and farmers' pesticide application behavior in China: A moderation and mediation model. *J. Clean. Prod.* **2020**, *276*, 124241. [CrossRef]
41. Duong, T.T.; Brewer, T.; Luck, J.; Zander, K. A Global Review of Farmers' Perceptions of Agricultural Risks and Risk Management Strategies. *Agriculture* **2019**, *9*, 10. [CrossRef]
42. Lu, X.; Zhang, Y.; Zou, Y. Evaluation the effect of cultivated land protection policies based on the cloud model: A case study of Xingning, China. *Ecol. Indic.* **2021**, *131*, 108247. [CrossRef]
43. Bar-Shira, Z.; Just, R.E.; Zilberman, D. Estimation of farmers' risk attitude: An econometric approach. *Agric. Econ.* **1997**, *17*, 211–222. [CrossRef]
44. Zeithaml, V.A. Consumer Perceptions of Price, Quality, and Value: A Means-End Model and Synthesis of Evidence. *J. Mark.* **1988**, *52*, 2–22. [CrossRef]
45. Nelson, A.C. Preserving Prime Farmland in the Face of Urbanization: Lessons from Oregon. *J. Am. Plan. Assoc.* **1992**, *58*, 467–488. [CrossRef]
46. Woodruff, R.B. Customer value: The next source for competitive advantage. *J. Acad. Mark. Sci.* **1997**, *25*, 139. [CrossRef]
47. Gai, H.; Yan, T.; Junbiao, Z. Perceived Value, Governament Regulations and Farmers' Behaviors of Continued Mechanized Operation of Straw Returning to the Field: An Analysis Based on Survery Data from 1288 Farmers in Three Provinces of Hebei, Anhui and Hubei. *Chin. Rural. Econ.* **2020**, *8*, 106–123, (In Chinese with English abstract).
48. Liu, H.; Luo, X. Understanding Farmers' Perceptions and Behaviors towards Farmland Quality Change in Northeast China: A Structural Equation Modeling Approach. *Sustainability* **2018**, *10*, 3345. [CrossRef]
49. Azadi, Y.; Yazdanpanah, M.; Mahmoudi, H. Understanding smallholder farmers' adaptation behaviors through climate change beliefs, risk perception, trust, and psychological distance: Evidence from wheat growers in Iran. *J. Environ. Manag.* **2019**, *250*, 109456. [CrossRef]
50. Liu, H.; Zhou, Y. Farmers' Cognition and Behavioral Response towards Cultivated Land Quality Protection in Northeast China. *Sustainability* **2018**, *10*, 1905. [CrossRef]
51. Scott, J.C. *The Moral Economy of the Peasant: Rebellion and Subsis-Tence in Southeast Asia*; Yale University Press: New Haven, CT, USA, 1976.
52. Zhou, Z.; Liu, J.; Zeng, H.; Zhang, T.; Chen, X. How does soil pollution risk perception affect farmers' pro-environmental behavior? The role of income level. *J. Environ. Manag.* **2020**, *270*, 110806. [CrossRef]
53. Lu, H.; Hu, L.; Zheng, W.; Yao, S.; Qian, L. Impact of household land endowment and environmental cognition on the willingness to implement straw incorporation in China. *J. Clean. Prod.* **2020**, *262*, 121479. [CrossRef]
54. Liu, Q.; He, J.; Hou, G. Farmers' Perceived Value and the New Rural Residential Satisfaction: The Regulatory Role of Peasant Household Characteristics. *Urban Dev. Stud.* **2013**, *20*, 96–101, (In Chinese with English abstract).
55. Agarwal, S.; Teas, R.K. Perceived Value: Mediating Role of Perceived Risk. *J. Mark. Theory Pract.* **2001**, *9*, 1–14. [CrossRef]
56. Sirivongs, K.; Tsuchiya, T. Relationship between local residents' perceptions, attitudes and participation towards national protected areas: A case study of Phou Khao Khouay National Protected Area, central Lao PDR. *For. Policy Econ.* **2012**, *21*, 92–100. [CrossRef]
57. Liao, W.; Xiang, D.; Chen, M.; Yu, J.; Luo, Q. The Impact of Perceived Value on Farmers' Regret Mood Tendency. *Sustainability* **2018**, *10*, 3650. [CrossRef]
58. Zhang, Y.; Xiao, C.; Zhou, G. Willingness to pay a price premium for energy-saving appliances: Role of perceived value and energy efficiency labeling. *J. Clean. Prod.* **2020**, *242*, 118555. [CrossRef]
59. Zhang, S.; Hu, W.; Li, M.; Guo, Z.; Wang, L.; Wu, L. Multiscale research on spatial supply-demand mismatches and synergic strategies of multifunctional cultivated land. *J. Environ. Manag.* **2021**, *299*, 113605. [CrossRef]
60. Jiang, G.; Wang, M.; Qu, Y.; Zhou, D.; Ma, W. Towards cultivated land multifunction assessment in China: Applying the "influencing factors-functions-products-demands" integrated framework. *Land Use Policy* **2020**, *99*, 104982. [CrossRef]
61. Xu, H.; Huang, X.; Zhong, T.; Chen, Z.; Yu, J. Chinese land policies and farmers' adoption of organic fertilizer for saline soils. *Land Use Policy* **2014**, *38*, 541–549. [CrossRef]
62. Menapace, L.; Colson, G.; Raffaelli, R. Risk Aversion, Subjective Beliefs, and Farmer Risk Management Strategies. *Am. J. Agric. Econ.* **2013**, *95*, 384–389. [CrossRef]
63. Fahad, S.; Wang, J.; Hu, G.; Wang, H.; Yang, X.; Shah, A.A.; Huong, N.T.L.; Bilal, A. Empirical analysis of factors influencing farmers crop insurance decisions in Pakistan: Evidence from Khyber Pakhtunkhwa province. *Land Use Policy* **2018**, *75*, 459–467. [CrossRef]
64. Ren, L.; Gan, C.; Wu, Y.; Chen, Y. Impacts of Farmers' Farmland Perceived Value on Farmers' Land Investment Behaviors in Urban Suburb: A Typical Sample Survey of Wuhan and Ezhou. *China Land Sci.* **2018**, *32*, 42–50. [CrossRef]

65. Akcaoz, H.; Ozkan, B. Determining risk sources and strategies among farmers of contrasting risk awareness: A case study for Cukurova region of Turkey. *J. Arid. Environ.* **2005**, *62*, 661–675. [CrossRef]
66. Tang, L.; Luo, X. Can agricultural insurance encourage farmers to apply biological pesticides? Evidence from rural China. *Food Policy* **2021**, *105*, 102174. [CrossRef]

sustainability MDPI

Article

Technification in Dairy Farms May Reconcile Habitat Conservation in a Brazilian Savanna Region

Juliana Silveira dos Santos [1,2], Fausto Miziara [3], Hayla da Silva Fernandes [3], Renato Cezar Miranda [4] and Rosane Garcia Collevatti [1,*]

1 Laboratório de Genética & Biodiversidade, Instituto de Ciências Biológicas, Universidade Federal de Goiás, Goiânia 74690-900, GO, Brazil; juliana.silveiradossantos@gmail.com
2 Departamento de Biodiversidade, Laboratório de Ecologia Espacial e Conservação (LEEC), Universidade Estadual Paulista Júlio de Mesquita Filho, UNESP, Rio Claro 13506-900, SP, Brazil
3 Programa de Pós-Graduação em Ciências Ambientais, Universidade Federal de Goiás, Goiânia 74690-900, GO, Brazil; faustomiziara@gmail.com (F.M.); hayla.fernandes@gmail.com (H.d.S.F.)
4 Floresta Nacional de Silvânia, Instituto Chico Mendes de Conservação da Biodiversidade (ICMBio), Silvânia 75180-000, GO, Brazil; renatocezarmiranda@gmail.com
* Correspondence: rosanegc68@icloud.com; Tel.: +55-62-3521-1729

Abstract: The assessment of the relationships between farm management systems and nature conservation may help in the design of more efficient strategies to uphold economic benefits and biodiversity conservation. To our knowledge, this is the first work in Brazil to study the relationship between farm conservation status and technification level. Here, we test the hypothesis that dairy farms with higher levels of technification have a higher percentage of natural vegetation and connectivity, and that differences in environment features between farms explain their conservation status. We obtained variables related to the level of technification such as feed, milking, sanitary control and breeding management systems. We show that farmers with a higher level of technification, such as artificial insemination in cattle breeding, tended to conserve a higher percentage of natural vegetation, as well as larger farms with a higher percentage of riparian forest. The adoption of artificial insemination is associated with other technification systems such as a forage diet, milking method and frequency and sanitary control. It is also significantly related to higher milk productivity. Our novel results point to a positive effect of technification on the conservation of natural vegetation, suggesting that economic incentives and programs aimed at increasing technification in cattle breeding may increase dairy production and conservation within the study area. Our findings also show an effect of larger areas of riparian forests, which are protected by Brazilian policy, in the conservation status of dairy farms.

Keywords: agricultural landscapes; Conefor; connectivity; legal reserve; long-term ecological research; spatial-temporal heterogeneity

Citation: dos Santos, J.S.; Miziara, F.; Fernandes, H.d.S.; Miranda, R.C.; Collevatti, R.G. Technification in Dairy Farms May Reconcile Habitat Conservation in a Brazilian Savanna Region. *Sustainability* **2021**, *13*, 5606. https://doi.org/10.3390/su13105606

Academic Editor: Yu Cao

Received: 5 April 2021
Accepted: 8 May 2021
Published: 17 May 2021

1. Introduction

The trade-off between food production and biodiversity conservation is one of the major challenges for policymakers in the Anthropocene [1]. The expansion of agriculture and the use of unsustainable practices are the main threats to terrestrial natural habitats [2] and to biodiversity conservation in tropical and temperate regions [3]. The global agricultural land area was 4.9 billion hectares (Gha) in 2016—38% of the global land surface [4]. Two thirds (3.3 Gha) of the agricultural land were used as pastures for livestock.

Pastures are the main anthropic disturbances world-wide and have expanded more rapidly than cropland. While agriculture expanded from 265 ha to 1471 million ha from 1700 to 1990, pastures expanded from 524 ha to 3451 million ha in the same period [5] and to 3340 million ha by 2005 [6]. This expansion is mainly due to the increasing pressure to produce animal protein, mainly from cattle. The global cattle livestock population was estimated to be 1.5 billion in 2012, producing 67 billion kg of beef carcass and 625 billion kg of milk [7]. The expected increase in animal protein demands of 1.3% per year until 2050 may lead to a 40% increase in cattle population [8].

Besides deforestation, livestock farming may cause high environmental impacts such as soil degradation [9,10], water eutrophication [11], water scarcity due to intense use and water spring deterioration [12]. In addition, livestock raising requires the use of fossil fuels in the entirety of the productive chain, leading to high emissions and pollution [13]. Methane is the main greenhouse gas in cattle raising and is responsible for 15% of global warming, representing ~14.5% (7.1 Gt CO_2 per year) of all anthropic emissions [14]. Livestock raising is largely viewed to be an unsustainable activity [15]. However, the direct impacts of livestock raising on biodiversity loss are still contentious. For instance, most studies in Europe show that livestock raising for meat or milk production in intensive farming systems has negative impacts, while livestock raising for land management or conservation has positive impacts [16]. In addition, the use of sustainable methods and technology to increase productivity has been suggested to increase and conserve on-farm biodiversity (e.g., [17,18]).

In Brazil, pastures for cattle livestock increased in the last 33 years from ~118 million ha $\pm$ 3.41% to ~178 million ha $\pm$ 2.53% [19]—an increase of nearly 60 million ha—leading to the rapid deforestation of Brazilian ecosystems. Presently, most of the pasture area is degraded (60%) or exhibits a reduced carrying capacity [20]. The increase of pasture productivity is of utmost importance to restrain the ongoing expansion in Brazil [21,22], using more technology in both cattle raising and pasture restoration. For instance, the restoration of degraded pastures in 2006 would have prevented the occupation of 147.5 million ha of new area in the Amazonia and Cerrado biomes [22].

The Cerrado biome in central–west Brazil is one of the major cattle raising regions, with 44% of the Brazilian cattle population and 60 million ha of pasture, which is the most dominant anthropic land use class [23]. From 2002 to 2013, the pasture area in the Cerrado increased by 11% and contributed to the consolidation of the anthropization of 50% of the biome [24], considered one of the world hotspots of biodiversity [25].

The expansion of cattle raising in the Brazilian Cerrado has been characterized by low levels of technification, such as free grazing and low pasture management, and has usually been associated with the primary land occupation [26]. As a consequence, degraded pastures currently represent 39% of Cerrado pastures [27]. Identifying the levels of technification—i.e., practices and management systems in feeding, pasture improvement, milking, sanitary control, stocking rates and animal selection and breeding—in livestock farming is a challenge in Brazil due to the wide variety of systems and technologies employed [28]. This diversity is mainly exhibited in feeding [29], milking management and breeding practices [30]. In Brazil, most dairy farmers still apply traditional production systems with low levels of technification, such as the natural service breeding method [31].

The relationship of the technification level in dairy farms and the conservation of natural vegetation is still overlooked in Brazil, especially in the Cerrado. Despite its huge biodiversity, the lack of information on how technification affects conservation limits sound conservation planning for this biome. The low productivity of pastures is a major driver of deforestation in Brazil, leading to overall environmental and socio-economic impacts [32,33]. Cattle ranching is mostly extensive and uses low-productivity systems, leading to pasture degradation [34]. However, although the adoption of technologies that improve production efficiency may also decrease impacts on natural resources and minimize greenhouse gas emissions, a rebound effect may lead to a loss of initial resource savings over time due to the increase in total resource use driven by socio-psychological adaptation ([35]; but see [36]). However, technification in pasture management in the Brazilian Amazonia, for instance, increased livestock and milk productivity and reduced environmental impacts such as soil degradation [37]. Because increasing the efficiency of livestock farming can reduce pasture expansion and deforestation [38], understanding how technification levels in farms can reduce deforestation is of utmost importance for Cerrado conservation and decreasing greenhouse gas emissions, as we investigate here.

Here we address the relationship between the level of technification and conservation in dairy farms in an intensive-farming landscape (Figure 1). We specifically test the hypothesis that dairy farms with higher levels of technification have higher amount of natural vegetation and connectivity. Alternatively, to account for the effects of environment, we test whether farm area and differences in environment features among farms, such as slope, percentage of riparian forest along water courses, percentage of agriculture and pasture explain conservation of natural vegetation in dairy farms. Riparian forests, i.e., forests adjacent to water courses, are Areas of Permanent Protection (APPs) meant to protect sensitive ecosystems by Brazilian environmental law. We hypothesize that farms with more water courses, and thus with higher amount of riparian forest and lower amount of agriculture or pasture have higher amount of natural vegetation and connectivity, despite the technification level. For this, we applied questionnaires to characterize dairy production and farm management system, and to obtain variables related to the level of technification in feed and milking management, sanitary control, and genetics and breeding systems. Using the Brazilian SICAR (Brazilian System of Rural Environment) database of rural properties, we obtained farm boundaries and mapped the area of natural vegetation remnants and estimated the connectivity among these remnants to predict the effects of technification on natural vegetation conservation.

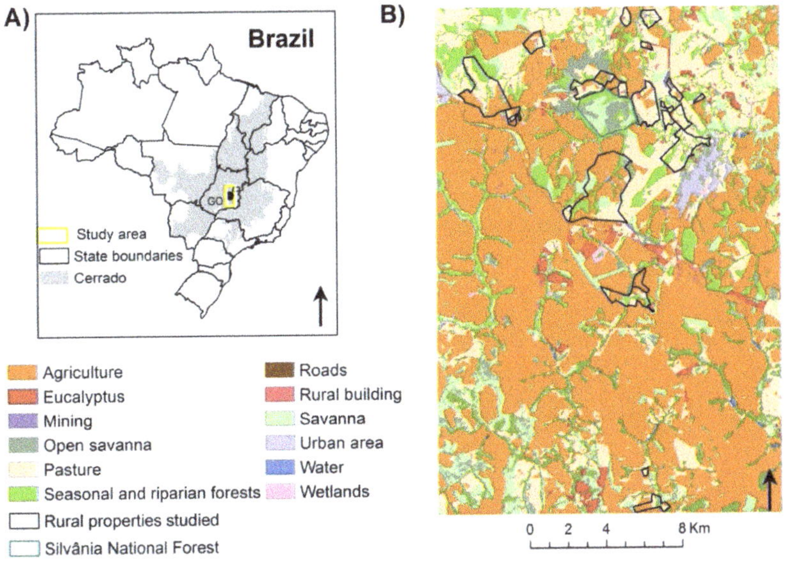

Figure 1. Spatial distribution of the 25 dairy farms in the COFA-LTER long-term project landscape. (**A**) Map of Brazil highlighting the study area. (**B**) Land use map of the COFA-LTER landscape highlighting the 25 dairy farms. Land use categories are in legends. The National Forest reserve is also highlighted in the map.

2. Materials and Methods

2.1. Study Area

The study was carried out in a Brazilian intensive-farming landscape comprising a Long-Term Ecological Research (LTER) project called COFA-LTER (Functional Connectivity in an Agricultural Landscape, Figure 1), in the Central-West Brazil, one of the most important Brazilian agribusiness regions (Figure 1). The COFA-LTER landscape comprises a reserve (Silvânia National Forest), the urban area of the city Silvânia, and the rural area with farms.

We mapped the land cover in COFA-LTER landscape using visual digitalization and manual classification of high-resolution images of Google Earth from 2019, freely available at the Geographic Information Systems QGis 2.4, with validation by field checking. The final map corresponded to 5 m spatial resolution, comprising 11 different land cover classes (Figure 1): (i) water courses; (ii) savanna and open savanna; (iii) seasonal and riparian forests; (iv) wetland; (v) pasture; (vi) agriculture (corn or soybean); (vii) rural building; (viii) mining; (ix) urban area; (x) road and train rail, and (xi) *Eucalyptus* spp. plantation.

All dairy farms in the COFA-LTER landscape (25 farms) were identified and mapped (Figure 1, Supplementary File S1, Table S1). Farm boundaries (Supplementary File S2, Figure S1) were obtained in the Brazilian SICAR (Brazilian System of Rural Environment) database of rural properties (http://www.car.gov.br/publico/imoveis/index, accessed on 15 February 2020) from 2019, or were manually digitalized in Google Earth.

We classified the farms according to the rural fiscal module of Silvânia (1 fiscal module = 30 ha; http://www.incra.gov.br/pt/, accessed on 15 February 2020): family farms < 30 ha); small size farms (1 to 4 modules); medium (4 to 15 modules); large (>15 modules).

2.2. Technification Variables and Farm Characteristics

We obtained variables related to the level of technification in the dairy farms and farm management practices using face to face interviews with farmers (25 farmers) in 2019. Before the interviews, we had several meetings with local stakeholders, including local government, non-governmental organizations (NGOs), researchers, farmers and technical assistants, to design a robust research approach in COFA-LTER project. All farmers signed an informed consent before interviewing.

We measured five different characteristics of technification in dairy production (Table 1), following the milk production systems prevalent in Brazil [39]: (i) productivity [40], measured by daily milk production; (ii) feed management [41,42], measured by primary forage diet, fertilizer pasture management, time of supplementary feed, criterium to supplement, and addition of vitamins to supplement; (iii) milking management [43], measured by milking method, milking frequency, and farm infrastructure for milking; (iv) sanitary control [44], measured by cleaning of udder before milking and CMT test (California Mastitis Test); (v) genetics and breeding [45], measured by breeding program for productivity improvement, selective breeding method and pregnancy diagnosis. We used daily milk production as a measure of productivity because farmers had no information on milk production per cow.

Table 1. Variables related to the level of technification and to overall milking management characteristics in dairy farms in COFA-LTER landscape.

Group	Variable	Possible Outcomes
Productivity	Daily milk production	Quantitative variable
Feed management	primary forage diet	(1) pasture-raised dairy (free grazing); (2) pasture-raised with rotational grazing; (3) semi-confinement; (4) confinement (compost Barn or free-stall)
	fertilizer pasture management	(1) yes (use fertilizers); (2) no (do not use)
	time of supplementary feed	(1) everyday; (2) dry season; (3) random; (0) not applicable (do not provide supplementary feed)
	criterium to supplement	(1) do not use; (2) technical calculation; (3) productivity; (4) random; (0) not applicable
	add vitamins to supplement	(1) yes; (2) no
Milking management	milking method	(1) full hand; (2) machine with bucket; (3) automatic machine
	milking frequency	(1) once-a-day; (2) twice-a-day; (3) three times a day
	farm infrastructure for milking	(1) milking pit; (2) milking parlor with roof and cement floor; (3) milking parlor without roof or cement floor; (4) milking parlor without roof and cement floor; (5) without a specific place to milk

Table 1. *Cont.*

Group	Variable	Possible Outcomes
Sanitary control	cleaning of udder before milking	(1) no cleaning; (2) water; (3) commercial sanitizer (pre-dipping)
	CMT test (California Mastitis Test)	(1) weekly; (2) biweekly; (3) monthly; (4) do not perform; (5) eventually with the suspicious of the disease
Genetics and breeding	breeding program for productivity improvement	(1) yes (perform breeding); (2) no
	selective breeding method	(1) natural service with selected breeding bulls; (2) artificial insemination; (0) do not perform selective breeding
	pregnancy diagnosis	(1) ultrasound; (2) transrectal palpation; (3) do not perform
Overall characteristics	Cattle breed composition	(1) holand; (2) gir; (3) girholand; (4) crossbred cattle
	Pasture restoration during the last decade	(1) yes (perform); (2) no (do not perform)
	Milk cooling tank	(1) bulk milk tank; (2) farm milk tank; (3) not applicable (do not use milk tank)
	Milk quality-based payment	(1) yes; (2) no
	Management separating pregnant cows	(1) yes; (2) no

Besides technification variables, we obtained variables related to the overall milking management to describe dairy farms in the studied area (Table 1): (i) cattle breed composition; (ii) pasture restoration during the last decade (iii) milk cooling tank; (iv) milk quality-based payment; (v) management separating pregnant cows.

2.3. Environment Feature Variables

We obtained the slope map from the TOPODATA INPE (National Institute for Space Research, http://www.dsr.inpe.br/topodata/acesso.php) database with 30 × 30 m spatial resolution. For each farm, we used the mean slope of all pixels. Using farm boundaries (Supplementary File S2, Figure S1), we identified areas of natural vegetation in each farm and calculated the farm area, the percentage of riparian forest, and percentage of agriculture and pasture. The available soil maps for the studied area were performed in small scale with few details (1:250,000) and show that the region is comprised mostly by dystrophic red oxisol with low variation among farms; thus, soil type could not be used as explanatory variable.

2.4. Conservation Status of the Dairy Farms

We used percentage of natural vegetation and connectivity as proxies of the conservation status of each farm (response variables). To verify the conservation status of natural vegetation that exceeds riparian forests in farms, we calculated the legal reserve area, which are areas of natural vegetation that should be preserved following Brazilian environmental law, corresponding to 20% of the total farm area in the Cerrado biome. To calculate legal reserve area, which is independent of APPs (Areas of Permanent Protection) such as riparian forests, we obtained the total percentage of natural vegetation and extracted the percentage of riparian forest. The highest slope in the 25 farms analyzed was ~13%, lower than the slope considered as APP by Brazilian environmental law (>45° in hillsides or 25° in hilltops). Thus, all APPs in our study area are riparian forests.

Connectivity is a measure of the landscape permeability to species movement among habitat patches [46,47]. To measure connectivity among remnant patches of natural vegetation in farms we used the Integral Index of Connectivity (IIC, [48]) implemented in Conefor [49] available in R version 3.6.1. [50]. The IIC, implemented in Conefor software [49], is based on the concept of graph theory and habitat availability, integrating habitat amount and the connectivity among patches of habitat [48]. IIC ranges from 0 (no connectivity among patches within the landscape) to 1 (high connectivity between remnant patches).

We calculated IIC considering each farm as a different landscape and habitat as all categories of natural vegetation. We used the mean value of IIC among four spatial distances as the response variable (100, 300, 500 and 1000 m), because most farms in the studied area were small and IIC values for all distances evaluated were very similar (Table S1).

2.5. Statistical Analyses

To analyze the effect of technification and environment features in conservation of dairy farms we first verified the multicollinearity among explanatory variables. For technification variables we used the generalized variance-inflation factor (Gvif), and for environment features we used the variance-inflation factor (Vif), excluding in this step the farm size which is a categorical variable. We used the *car* package [51] implemented in R version 3.6.1. For technification, we ran one model per variable group (Table 1). We used a stepwise approach eliminating the models with Gvif or Vif > 5.0 [52]. Gvif is equivalent to Vif for categorical variables, the inflation in size of the confidence ellipse or ellipsoid for the coefficients of the predictor variable in comparison with what would be obtained for orthogonal, uncorrelated data [53].

We then analyzed the effects of technification and environment features on farm conservation status using linear models. We defined all categorical variables as "factors," except daily milk production and environment variables, which correspond to continuous variables. To select the best variable explaining percentage of natural vegetation or IIC, we used the function *drop1* in R version 3.6.1. [50] with stepwise backward selection and Chi-Square distribution to calculate the significance of each variable per group, considering technification and environment variables separately. We considered as significant all variables with $p \leq 0.05$, and marginally significant variables with $p \leq 0.10$.

We tested the significance of association between pairs of technification variables using Kendall's coefficient of concordance (Kendall's tau-b) to test for concordance or attribute agreement between variables. Analyses were performed using the software Minitab®18.

3. Results

3.1. Dairy Farm Characteristics

Although farmers declared legal reserve area equal to the minimum required by Brazilian environmental law in Central-Brazil (20% of the farm area, Table S1), 84% of the farms had lower percentage of natural vegetation than the minimum required for legal reserve (Figure 2a, Supplementary Table S1). The total area of natural vegetation (Figure S2, Table S1) comprised 5.42 to 46.55% of the farm area (mean = 24.84%, SD = 11.45%), and only 16% of the dairy farms had proportion of natural vegetation > 20%, excluding riparian forests (Figure 2a).

Most dairy farms (80%) were family or small size, and only one was a large dairy farm (Figure 2b, Table S1, Figure S2). Mean IIC ranged from 0.0023 to 0.2141 (Figure S3, Table S1), and the mean overall farms was very low, 0.0646 (SD = 0.0561), meaning that the connectivity among remaining patches of natural vegetation within farms was very low (Figure 2c).

Most dairy farms use free grazing (pasture-raised dairy) and semi-confinement as primary forage diet system (Table S2). They use fertilizers in pasture management and perform pasture restoration, and provide supplementary feed during the dry season (Table S2). Most farmers use productivity as the criterium to give supplement to the cattle. Machine with bucket is the most used milking method, and milking twice-a-day is more frequently used (Table S2). Most farmers milk in a milking pit or in a milking parlor with roof and cement floor, use farm milk cooling tank and can receive milk quality-based payment (Table S2). Most farmers use commercial sanitizer to cleaning udder before milking and do not perform CMT test or perform monthly (Table S2). Most farmers use selective breeding method with natural service with selected breeding bulls and pregnancy diagnosis with transrectal palpation (Table S2). Farmers have Holland or crossbreed dairy cattle and do not separate pregnant cows.

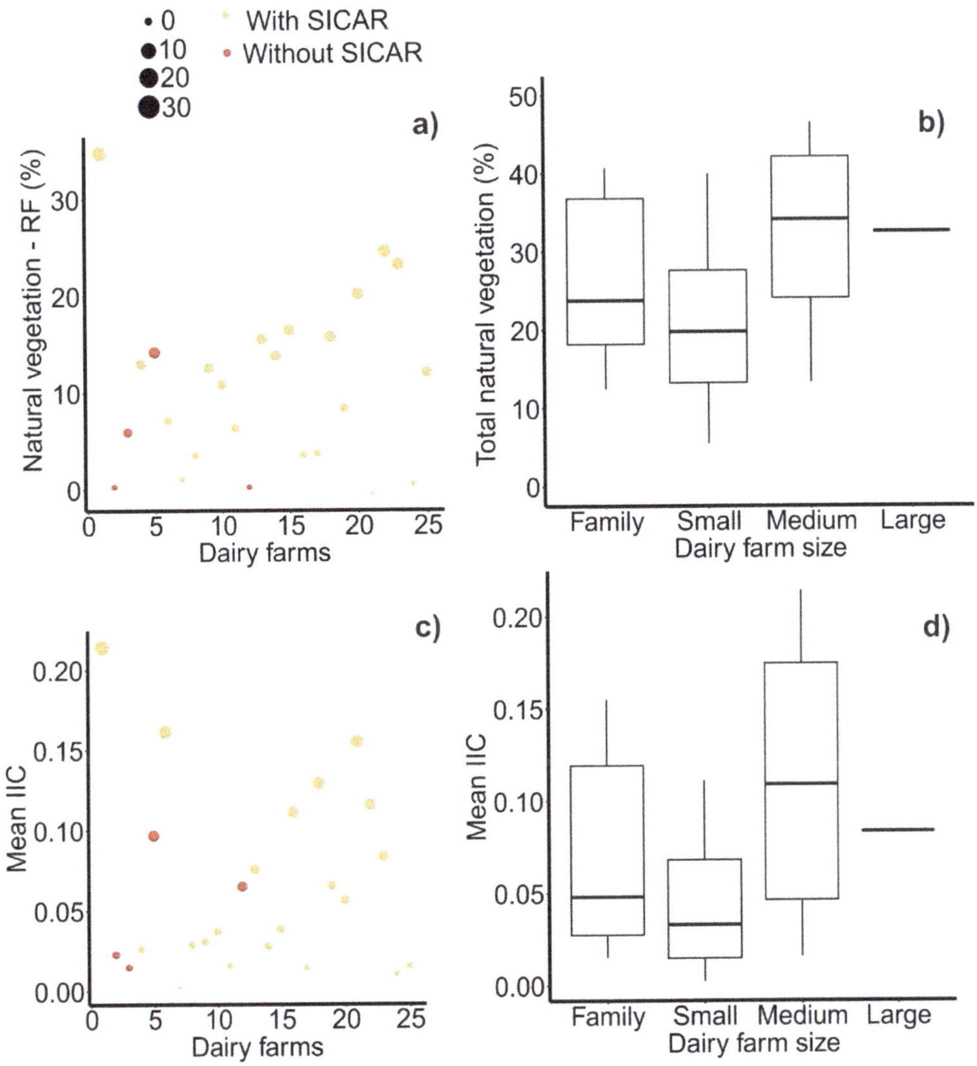

Figure 2. Percentage of natural vegetation and connectivity in the dairy farms of the COFA-LTER long-term project landscape. (**a**) Distribution of the percentage of legal reserve (total natural vegetation-riparian forest (RF)) in dairy farms with (dark red circles) and without SICAR (Brazilian System of Rural Environment, CAR, dark yellow circles). (**b**) Relationship between the total percentage of natural vegetation and dairy farm size (fiscal module). (**c**) Distribution of the Integral Index of Connectivity (IIC) in dairy farms with and without SICAR. (**d**) Relationship between IIC and dairy farm size. The box plot represents the median (dark bar), the third quartiles (the box) and the minimum (lower bar) and maximum (above bar) values.

3.2. Technification and Conservation Status

We excluded the variables primary forage diet (Gvif = 8.974) for feed management (Table S3), and milking method (Gvif = 5.093) for milking management (Table S3) due to collinearity. For sanitary control and genetics and breeding we kept all variables (Gvif < 5.0, Table S3).

We found that selective breeding method ($p = 0.046$) significantly explained the variation in percentage of natural vegetation among dairy farms (Tables 2 and S4). Dairy farms applying artificial insemination (Figure 3a) had higher proportion of natural vegetation. Breeding program for productivity improvement was marginally significant ($p = 0.059$) and daily milk production did not explain variation in percentage of natural vegetation ($p = 0.789$). When analyzed in the final model (Table 2), feed and milking management, and sanitary control variables did not explain farm conservation status (all $p > 0.05$; Table S4). Connectivity (mean IIC) was not explained by feed and milking management or sanitary control variables, or daily milk production (all $p > 0.05$; Table S5).

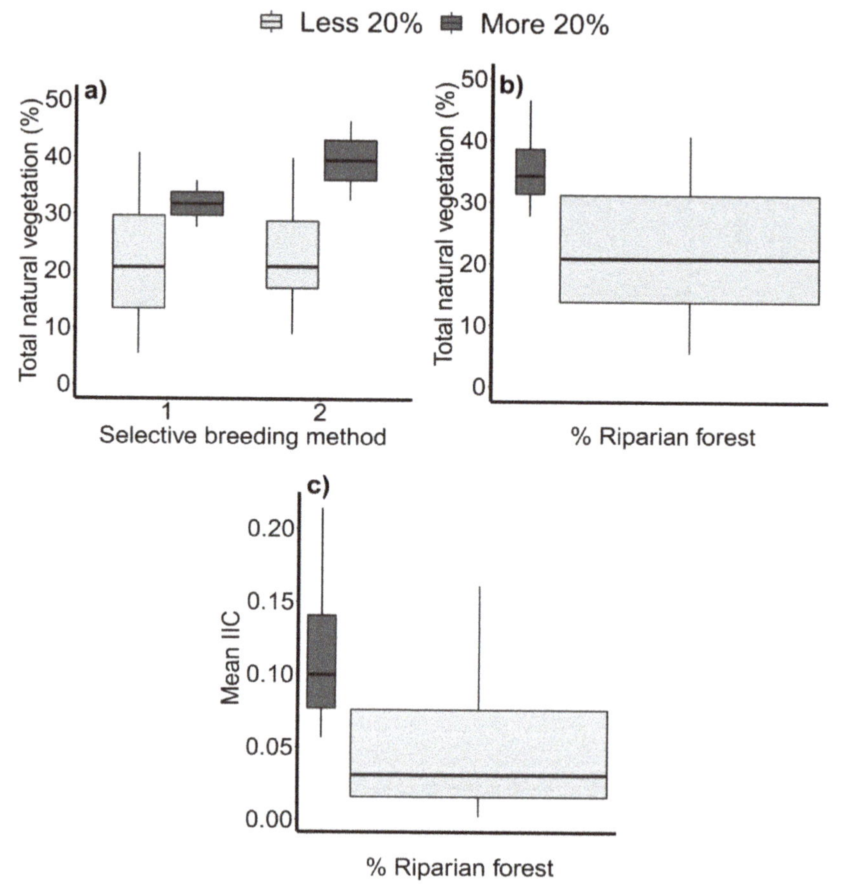

Figure 3. Relationship between conservation status and technification variable related to genetics and breeding, and environment feature in the 25 dairy farms in the COFA-LTER long-term project landscape. The relationships are shown for farms with less (light gray) and higher (dark gray) minimum percentage of legal reserve required by Brazilian law (20%). (**a**) Percentage of natural vegetation and selective breeding method. (**b**) Percentage of natural vegetation and riparian forest. (**c**) Integral Index of Connectivity (IIC) and percentage of riparian forest. 1, natural service with selected breeding bulls; 2, artificial insemination. The box plot represents the median (dark bar), the third quartiles (the box) and the minimum (lower bar) and maximum (above bar) values.

Table 2. Relationship between technification variables and conservation status measured by the percentage of natural vegetation in dairy farms in COFA-LTER landscape. Df, degrees of freedom, RSS, residual sum of square, AIC, Akaike Information Criterion.

Model	df	Sum of Square	RSS	AIC	*p*
Breeding program for productivity improvement	1	395.86	2996.4	125.66	0.059 *
Selective breeding method	1	448.57	3049.1	126.09	0.046 **
Daily milk production	1	7.49	2608.1	122.19	0.789

** significant ($p \leq 0.05$), * marginally significant ($p \leq 0.10$).

We found significant association between selective breeding method and primary forage diet ($p = 0.001$, Table S6). In addition, daily milk production was significantly higher ($t = 2.73$, $p = 0.026$, Figure 4) in farms that used artificial insemination (mean production = 1023 l, SD = 933.0 l) than in farms that used natural service with selected breeding bulls (mean production = 166.0, SD = 165.0).

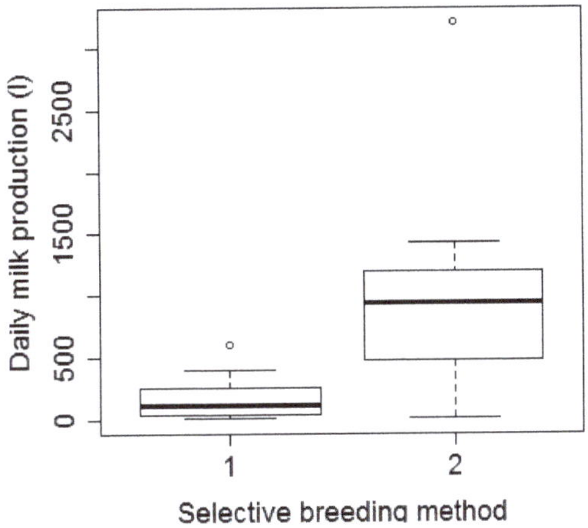

Figure 4. Relationship between selective breeding method and daily milk production in the 25 dairy farms in the COFA-LTER long-term project landscape. 1, natural service with selected breeding bulls; 2, artificial insemination. The box plot represents the median (dark bar), the third quartiles (the box) and the minimum (lower bar) and maximum (above bar) values.

3.3. Environment Features and Conservation Status

We excluded percentage of agriculture (Vif = 7.945) due to collinearity (Table S7), keeping slope, percentage of riparian forest and pasture in the models (Vif < 5.0, Table S7).

Percentage of riparian forest ($p < 0.001$) and farm size ($p = 0.041$) significantly explained the percentage of natural vegetation conserved in dairy farms (Table 3, Table S8). Farms with higher percentage of riparian forest (Figure 3b) or larger farms (Figure 2b) had higher percentage of natural vegetation. Slope and percentage of pasture did not explain the percentage of natural vegetation conserved in dairy farms (all $p > 0.05$, Table S8). We found similar results for IIC (Table 3, Table S9). Farms with higher percentage of riparian forest ($p < 0.001$, Figure 3c) and larger farms ($p = 0.019$, Figure 2d) had significantly higher connectivity.

Table 3. Relationship between environment feature variables and conservation status measured by the percentage of natural vegetation and the Integral Index of Connectivity (IIC) in dairy farms in COFA-LTER landscape. Df, degrees of freedom, RSS, residual sum of square, AIC, Akaike Information Criterion.

Model	df	Sum of Square	RSS	AIC	p
Percentage of natural vegetation					
Percentage of riparian forest	1	1420.24	2652.5	124.61	1.198×10^{-5} **
Size of rural property	3	483.08	1715.3	109.71	0.041 **
IIC					
Percentage of riparian forest	1	0.0314	0.0603	−142.66	1.851×10^{-5} *
Size of rural property	3	0.0139	0.0429	−155.16	0.019 *

** significant ($p \leq 0.05$), * marginally significant ($p \leq 0.10$).

4. Discussion

Here we show for the first time that farmers using cutting edge technology such as artificial insemination in cattle breeding tend to conserve higher percentage of natural vegetation in the COFA-LTER landscape. However, larger dairy farms and with higher percentage of riparian forest also showed higher conservation status. Cattle raising farms in Brazil use predominantly extensive pasture-raised (free grazing) system with low productivity, which may hamper conservation of natural vegetation due to the constant need for new pasture areas [21,54]. In the Cerrado biome, ~39% of pastures are degraded [27]. In the COFA-LTER landscape, most dairy farmers use pasture-raised with free grazing or semi-confinement that includes free grazing and free-stall or Barn systems, which is similar to the most used feed management system in Brazilian dairy farms (e.g., [55]). The predominance of these feed management systems may contribute to the relatively low percentage of natural vegetation and connectivity among patches of vegetation in dairy farms due to the demand of pasture and grain plantation for cattle feeding. Few dairy farmers (16%) had higher proportion of vegetation remnants than 20%, which is the minimum determined by Brazilian environmental law (see Figure 2a). Moreover, larger farms with higher percentage of riparian forest preserved higher percentage of natural vegetation, and had higher connectivity (see Figure 2b,c). However, it is important to note that overall connectivity was very low, which may compromise biodiversity conservation in the landscape.

Indeed, this result was not a surprise, because farmers do not have technical support to design priority areas for conservation within the rural property that would maximize the trade-offs between ecological and economic benefits, such as improving connectivity. Usually, farmers in Brazil conserve unproductive areas, or with technical limitations for management that will not interfere in daily management practices (see [56]). The higher conservation status in larger farms with higher percentage of riparian forest suggests that in the COFA-LTER landscape farmers are not preserving vegetation above the required by Brazilian environmental law. Farmers may adopt conservation practices based mainly on short economic benefits and in other factors such as previous experience, familiarity with technologies, perceived risks, labor requirements, and interactions with peers and advisors [57]. In addition, public visibility and the influence of neighbor farmers that use conservation practices can drive the decision of surrounding farmers to adopt new or more ecological management practices [57]. To our knowledge, the local government has no economic incentives to guarantee conservation, restoration or sustainable practices to improve local and regional connectivity and conservation.

Furthermore, the adherence of farmers to the SICAR (Brazilian System of Rural Environment) seems to be more related to restrictions to credit and rural financing and insurance imposed by Brazilian policies than to conservation. SICAR declaration is related to the access to rural credit and adoption of practices to improve pasture management in cattle farms in Brazil [32]. In the COFA-LTER four dairy farms have no register in the Brazilian government database SICAR, that requires the information of legal reserve area or which area will be restore in the future to comprise the legal reserve and APPs. However, even with the SICAR declaration, most of the dairy farmers in the COFA-LTER landscape (84% of the dairy farms) do not conserve the minimum of natural vegetation established by law, suggesting that the current environmental policies provide no additional incentive to conservation in the study area.

However, it is important to highlight that positive outcomes can be identified at regional scales, as we found higher percentage of natural vegetation in a set of dairy farms that use higher levels of technification (see Figure 3). Furthermore, planned interventions with few interferences on the production systems can favor local biodiversity and ecosystem services [58,59]. Some of these interventions support the restoration of strips of natural vegetation between the crop fields [60] and the reduction of agrochemical inputs [61]. In the COFA-LTER landscape farmers are prone to restore areas of natural vegetation, particularly farmers with ecosystem service awareness and with higher number of springs in their properties [62].

Feed management may affect greenhouse gas emission directly due to both the forage production and feed conversion rates [63], and indirectly due to expansion of new areas of pasture and grain plantation for animal feed leading to deforestation [64,65]. However, we found no relationship between feed management and the percentage of natural vegetation, and also no relationship of percentage of pasture and conservation status. It is possible that farmers may have cleared their farms for pasture planting before the adoption of cutting-edge technology, and then turned to higher levels of technification [21,32,66] such as insemination, milking using automatic machines and confinement.

The increase in productivity of cattle raising and in general, the ecological intensification of agriculture [57] may improve not only direct economic incomes but also conservation of vegetation remnants [22]. In fact, our results evince that increasing technology and cattle raising efficiency may improve economic benefits. Artificial insemination was directly related to productivity in COFA-LTER landscape, i.e., dairy farms applying artificial insemination tended to have higher daily milk production (see Figure 4), which may increase economic benefits. The benefits of applying technification can also be detected in other important indicators of productivity, for instance, dairy farms adopting artificial insemination (mean = 48.6, SD = 30.5) have higher number of dairy cows ($t = 3.25$, $p = 0.012$) than those with natural service with selected breeding bulls (mean = 14.9, SD = 7.4).

However, although our novel findings indicate opportunities to improve economic and ecological benefits on dairy farms, these opportunities may vary throughout Brazilian territory, restricting our conclusions to the study area. Also, it is not easy to establish patterns in the production systems in Brazil. The country is very heterogeneous in a set of factors such as the environment, technology availability and adoption, which may establish patterns utterly different in the same production system such as dairy farms.

Sustainable intensification in Brazil can potentially increase 113% in pasture grazing beef and milk production without increasing land demand and sparing areas for biodiversity conservation [67]. Our findings point to important outcomes about technification and conservation in the COFA-LTER landscape, showing that additional economic incentives and programs aimed at improving technification in cattle breeding can increase dairy production and, consequently, may lead to additional incentives to the conservation and restoration of vegetation remnants. Besides artificial insemination, our results suggest that practices such as confinement and semi-confinement may contribute to less land demand. However, it is essential to mention that non-ecological intensification practices may reduce pasture area, but may cause a set of additional impacts on the environment. Farmers tend to adopt new management practices when they do not require high modification in the establish farming systems [57]. Furthermore, evidence-based new management systems have higher chances of adoption when the most relevant costs and benefits to the farmers are clear [57]. In this context, we are providing evidence that technification may bring positive income to dairy farms in the COFA-LTER region.

5. Concluding Remarks

This is the first study in Brazil showing that dairy farms using cutting edge technology in cattle breeding, such as artificial insemination, significantly conserve higher proportion of natural vegetation in their farms. Therefore, technification affected positively the conservation of natural vegetation in the COFA-LTER landscape. Farmers adopting artificial insemination had higher daily milk production and tended to use other technification systems related to dairy farm management more frequently than the other farmers, such as confinement and semi-confinement, and milking with automatic machine.

Brazil harbors a diversity of social, environmental, and economic conditions with can provide different perspectives to other parts of the territory and other productive systems (see an example in [32]). Our dataset provides evidence that incentives to technification may promote economic benefits to dairy farmers and can provide opportunities to preserve or restore habitats in COFA-LTER landscape. Our dataset can be a starting point to design successful strategies aimed at the conservation and dairy production in the study area. However, the use of new practices or strategies can be seen as a barrier if enable conditions are not offered safely, like technical assistance and training for labor [32,67].

The significant relationships between farm size and percentage of riparian forest and farm conservation status highlight the importance of riparian forests to maintain the connectivity at farm level, and to favor conservation status in productive dairy regions. Ecological initiatives should be designed at local and regional levels to favor essential components such as connectivity and designing farms and landscapes more permeable to organism's movement. Additional incentives of local government are essential to support the dairy farmers to achieve economic benefits, sustainable intensification, maintain or restore habitats, and ensuring ecological benefits within farms. These initiatives are necessary independently of farm size, to promote a conservation status at landscape level. We acknowledge the limitations of our results due to the sampling size, i.e., the number of dairy farms analyzed, and the dairy farm sizes. Most farms in the COFA-LTER region are family, small or medium size, thus our results may be used with caution to large size and very large states. Also, the specific environmental conditions of the region may also influence the results, since riparian forests are important landscape components in the region due to the topography and geomorphology. Also, more studies are necessary at the farm level to understand the relationship between technification and conservation, identify barriers, and promote efficient strategies to fix it.

Supplementary Materials: The following are available online at https://www.mdpi.com/article/10.3390/su13105606/s1, Supplementary File S1: Tables S1–S9, Supplementary File S2: Figures S1–S3, Video S1: video. abstract. Supplementary S1 Tables (Tables S1–S9) with raw data and results of IIC, Gvif, Vif and GLM models. Supplementary S2 Figures (Figures S1–S3) with the spatial distribution of the 25 dairy farms in the COFA-LTER long-term project landscape, distribution of dairy farms size and natural vegetation and distribution of IIC.

Author Contributions: Conceptualization, F.M., R.G.C. and H.d.S.F.; methodology, F.M., R.G.C., J.S.d.S. and H.d.S.F.; supervision, R.G.C. and F.M.; validation, R.G.C. and F.M.; investigation, H.d.S.F. and R.C.M., formal analysis, J.S.d.S. and R.G.C. data curation, R.G.C.; original draft preparation R.G.C. and J.S.d.S.; writing review and editing, R.G.C. and J.S.d.S.; funding acquisition, R.G.C. All authors have read and agreed to the published version of the manuscript.

Funding: This work was supported by grants to the research network COFA-LTER supported by MCT/CNPq/CAPES (project no 441278/2016-7) and FAPEG (project no. 201710267000331), CAPES/PROCAD (project no. 88881.068425/2014-01). JSS received a CAPES postdoctoral fellowship and a postdoctoral grant from São Paulo Research Foundation (FAPESP, process 2019/09713-6).

Institutional Review Board Statement: The statistical analyses involved only analyses of the rural property data and no association between farmer's personal data and the rural property data. The wok was approved by the Universidade Federal de Goiás (UFG, Federal University of Goiás, Brazil) ethics committee.

Informed Consent Statement: The present work involved face to face interviews with farmers. All farmers signed an informed consent for before being interviewed, as stated by Brazilian law.

Data Availability Statement: Data and additional supporting information may be found in the online version of this article as supporting information.

Acknowledgments: R.G.C. has continuously been supported by productivity grants from CNPq and Capes, which we gratefully acknowledge. We thank two anonymous reviewers for helpful comments to improve the manuscript.

Conflicts of Interest: The authors declare that the research was conducted in the absence of any commercial, financial or non-financial relationships that could be construed as a potential conflict of interest.

References

1. Johnson, C.N.; Balmford, A.; Brook, B.W.; Buettel, J.C.; Galetti, M.; Guangchun, L.; Wilmshurst, J.M. Biodiversity losses and conservation responses in the Anthropocene. *Science* **2017**, *356*, 270–275. [CrossRef]
2. Laurance, W.F.; Sayer, J.; Cassman, K.G. Agricultural expansion and its impacts on tropical nature. *Trends Ecol. Evol.* **2014**, *29*, 107–116. [CrossRef]
3. Newbold, T.; Hudson, L.N.; Hill, S.L.L.; Contu, S.; Lysenko, I.; Senior, R.A.; Börger, L.; Bennett, D.J.; Choimes, A.; Collen, B.; et al. Global effects of land use on local terrestrial biodiversity. *Nature* **2015**, *520*, 45–50. [CrossRef]
4. FAO. *The State of the World's Biodiversity for Food and Agriculture*; Bélanger, J., Pilling, D., Eds.; FAO: Rome, Italy, 2019; ISBN 978-92-5-131270-4.
5. Goldewijk, K.K. Estimating global land use change over the past 300 years: The HYDE database. *Glob. Biogeochem. Cycles* **2001**, *15*, 417–433. [CrossRef]
6. Hurtt, G.C.; Chini, L.P.; Frolking, S.; Betts, R.A.; Feddema, J.; Fischer, G.; Fisk, J.P.; Hibbard, K.; Houghton, R.A.; Janetos, A.; et al. Harmonization of land-use scenarios for the period 1500–2100: 600 years of global gridded annual land-use transitions, wood harvest, and resulting secondary lands. *Clim. Chang.* **2011**, *109*, 117–161. [CrossRef]
7. LEAP Environmental Performance of Large Ruminant Supply Chains: Guidelines for Assessment. Draft for Public Review. Livestock Environmental Assessment and Performance (LEAP) Partnership. Available online: http://www.fao.org/3/a-bl094e.pdf (accessed on 17 May 2021).
8. Alexandratos, N.; Bruinsma, J. World Agriculture towards 2030/2050: The 2012 Revision. Available online: http://www.fao.org/3/a-ap106e.pdf (accessed on 17 May 2021).
9. Clegg, C.D. Impact of cattle grazing and inorganic fertiliser additions to managed grasslands on the microbial community composition of soils. *Appl. Soil Ecol.* **2006**, *31*, 73–82. [CrossRef]
10. Kolbek, J.; Alves, R.J.V. Impacts of cattle, fire and wind in rocky savannas, southeastern Brazil. *Acta Univ. Carolinae Environ.* **2008**, *22*, 111–130.
11. Burt, C.; Bachoon, D.S.; Manoylov, K.; Smith, M. The impact of cattle farming best management practices on surface water nutrient concentrations, faecal bacteria and algal dominance in the Lake Oconee watershed. *Water Environ. J.* **2012**, *27*, 207–215. [CrossRef]
12. Conroy, E.; Turner, J.N.; Rymszewicz, A.; O'Sullivan, J.J.; Bruen, M.; Lawler, D.; Lally, H.; Kelly-Quinn, M. The impact of cattle access on ecological water quality in streams: Examples from agricultural catchments within Ireland. *Sci. Total Environ.* **2016**, *547*, 17–29. [CrossRef]
13. De Boer, I.J.M.; Cederberg, C.; Eady, S.; Gollnow, S.; Kristensen, T.; Macleod, M.; Meul, M.; Nemecek, T.; Phong, L.T.; Thoma, G.; et al. Greenhouse gas mitigation in animal production: Towards an integrated life cycle sustainability assessment. *Curr. Opin. Environ. Sustain.* **2011**, *3*, 423–431. [CrossRef]
14. Gerber, P.J.; Steinfeld, H.; Henderson, B.; Mottet, A.; Opio, C.; Dijkman, J.; Falcucci, A.; Tempio, G. *Tackling Climate Change through Livestock—A Global Assessment of Emissions and Mitigation Opportunities*; FAO: Rome, Italy, 2013.
15. Perfecto, I.; Vandermeer, J. Biodiversity conservation in tropical agroecosystems: A new conservation paradigm. *Ann. N. Y. Acad. Sci.* **2008**, *1134*, 173–200. [CrossRef]
16. Kok, A.; de Olde, E.M.; de Boer, I.J.M.; Ripoll-Bosch, R. European biodiversity assessments in livestock science: A review of research characteristics and indicators. *Ecol. Indic.* **2020**, *112*, 105902. [CrossRef]
17. Alkemade, R.; Reid, R.S.; Van Den Berg, M.; De Leeuw, J.; Jeuken, M. Assessing the impacts of livestock production on biodiversity in rangeland ecosystems. *Proc. Natl. Acad. Sci. USA* **2013**, *110*, 20900–20905. [CrossRef]
18. Broom, D.M.; Galindo, F.A.; Murgueitio, E. Sustainable, efficient livestock production with high biodiversity and good welfare for animals. *Proc. R. Soc. B Biol. Sci.* **2013**, *280*. [CrossRef]
19. Parente, L.; Mesquita, V.; Miziara, F.; Baumann, L.; Ferreira, L. Assessing the pasturelands and livestock dynamics in Brazil, from 1985 to 2017: A novel approach based on high spatial resolution imagery and Google Earth Engine cloud computing. *Remote Sens. Environ.* **2019**, *232*, 111301. [CrossRef]

20. Nogueira, S.F.; Macedo, M.C.M.; Poccard-Chapuis, R.; Da Silva, G.B.S.; Victoria, D.C.; Andrade, R.G.; Aguiar, D.A.; Bolfe, E.L. Geotechnologies for monitoring pasture degradation levels in Brazil. In *A Produção Animal Frente às Mudanças Climáticas e Tecnológicas, Proceedings of the 51 Reunião Anual Da Sociedade Brasileira De Zootecnia, Barra dos Coqueiros, Brazil, 19 December 2014*; Sociedade Brasileira de Zootecnia: Barra dos Coqueiros, Brazil, 2014.
21. Latawiec, A.E.; Strassburg, B.B.N.; Valentim, J.F.; Ramos, F.; Alves-Pinto, H.N. Intensification of cattle ranching production systems: Socioeconomic and environmental synergies and risks in Brazil. *Animal* **2014**, *8*, 1255–1263. [CrossRef]
22. Strassburg, B.B.N.; Latawiec, A.E.; Barioni, L.G.; Nobre, C.A.; da Silva, V.P.; Valentim, J.F.; Vianna, M.; Assad, E.D. When enough should be enough: Improving the use of current agricultural lands could meet production demands and spare natural habitats in Brazil. *Glob. Environ. Chang.* **2014**, *28*, 84–97. [CrossRef]
23. Scaramuzza, C.A.d.M.; Sano, E.E.; Adami, M.; Bolfe, E.L.; Coutinho, A.C.; Esquerdo, J.C.D.M.; Maurano, L.E.P.; Narvaes, I.S.; Oliveira Filho, F.J.B.; Rosa, R.; et al. Land-Use and Land-Cover Mapping of the Brazilian Cerrado Based Mainly on Landsat-8 Satellite Images. *Rev. Bras. Cartogr.* **2017**, *69*, 1041–1051.
24. Sano, E.E.; Rosa, R.; Scaramuzza, C.A.d.M.; Adami, M.; Bolfe, E.L.; Coutinho, A.C.; Esquerdo, J.C.D.M.; Maurano, L.E.P.; Narvaes, I.d.S.; Filho, F.J.B.d.O.; et al. Land use dynamics in the Brazilian Cerrado in the period from 2002 to 2013. *Pesqui. Agropecuária Bras.* **2019**, *54*. [CrossRef]
25. Myers, N.; Mittermeier, R.A.; Mittermeier, C.G.; da Fonseca, G.A.B.; Kent, J. Biodiversity hotspots for conservation priorities. *Nature* **2000**, *403*, 853–858. [CrossRef] [PubMed]
26. De Sy, V.; Herold, M.; Achard, F.; Beuchle, R.; Clevers, J.G.P.W.; Lindquist, E.; Verchot, L. Land use patterns and related carbon losses following deforestation in South America. *Environ. Res. Lett.* **2015**, *10*. [CrossRef]
27. Pereira, O.J.R.; Ferreira, L.G.; Pinto, F.; Baumgarten, L. Assessing pasture degradation in the Brazilian Cerrado based on the analysis of MODIS NDVI time-series. *Remote Sens.* **2018**, *10*, 1761. [CrossRef]
28. De Oliveira Silva, R.; Barioni, L.G.; Queiroz Pellegrino, G.; Moran, D. The role of agricultural intensification in Brazil's Nationally Determined Contribution on emissions mitigation. *Agric. Syst.* **2018**, *161*, 102–112. [CrossRef]
29. Oliveira, C.A.; Millen, D.D. Survey of the nutritional recommendations and management practices adopted by feedlot cattle nutritionists in Brazil. *Anim. Feed Sci. Technol.* **2014**, *197*, 64–75. [CrossRef]
30. Costa, J.H.C.; Hötzel, M.J.; Longo, C.; Balcão, L.F. A survey of management practices that influence production and welfare of dairy cattle on family farms in southern Brazil. *J. Dairy Sci.* **2013**, *96*, 307–317. [CrossRef] [PubMed]
31. Balcão, L.F.; Longo, C.; Costa, J.H.C.; Uller-Gómez, C.; Filho, L.C.P.M.H.; Hötzel, M.J. Characterisation of smallholding dairy farms in southern Brazil. *Anim. Prod. Sci.* **2017**, *57*, 735–745. [CrossRef]
32. Latawiec, A.E.; Strassburg, B.B.N.; Silva, D.; Alves-Pinto, H.N.; Feltran-Barbieri, R.; Castro, A.; Iribarrem, A.; Rangel, M.C.; Kalif, K.A.B.; Gardner, T.; et al. Improving land management in Brazil: A perspective from producers. *Agric. Ecosyst. Environ.* **2017**, *240*, 276–286. [CrossRef]
33. Latrubesse, E.M.; Arima, E.; Ferreira, M.E.; Nogueira, S.H.; Wittmann, F.; Dias, M.S.; Dagosta, F.C.P.; Bayer, M. Fostering water resource governance and conservation in the Brazilian Cerrado biome. *Conserv. Sci. Pract.* **2019**, *1*, 1–8. [CrossRef]
34. Dias-Filho, M.B. Diagnóstico das Pastagens no Brasil. *Embrapa Amaz. Orient.* **2014**, *22*. [CrossRef]
35. Gillingham, K.; Rapson, D.; Wagner, G. The rebound effect and energy efficiency policy. *Rev. Environ. Econ. Policy* **2016**, *10*, 68–88. [CrossRef]
36. Paul, C.; Techen, A.K.; Robinson, J.S.; Helming, K. Rebound effects in agricultural land and soil management: Review and analytical framework. *J. Clean. Prod.* **2019**, *227*, 1054–1067. [CrossRef]
37. zu Ermgassen, E.K.H.J.; de Alcântara, M.P.; Balmford, A.; Barioni, L.; Neto, F.B.; Bettarello, M.M.F.; de Brito, G.; Carrero, G.C.; Florence, E.d.A.S.; Garcia, E.; et al. Results from on-the-ground efforts to promote sustainable cattle ranching in the Brazilian Amazon. *Sustainability* **2018**, *10*, 1301. [CrossRef]
38. Steinfeld, H.; Wassenaar, T.; Jutzi, S. Livestock production systems in developing countries: Status, drivers, trends. *Rev. Sci. Tech. l'OIE* **2006**, *25*, 505–516. [CrossRef] [PubMed]
39. Neto, M.C.; Campos, J.M.d.S.; de Oliveira, A.S.; Gomes, S.T. Identification and quantification of benchmarks of milk production systems in Minas Gerais1. *Rev. Bras. Zootec.* **2012**, *41*, 2279–2288. [CrossRef]
40. de Andrade Ferrazza, R.; Lopes, M.A.; de Oliveira Prado, D.G.; de Lima, R.R.; Bruhn, F.R.P. Association between technical and economic performance indexes and dairy farm profitability. *Rev. Bras. Zootec.* **2020**, *49*. [CrossRef]
41. Bargo, F.; Muller, L.D.; Delahoy, J.E.; Cassidy, T.W. Performance of high producing dairy cows with three different feeding systems combining pasture and total mixed rations. *J. Dairy Sci.* **2002**, *85*, 2948–2963. [CrossRef]
42. Sairanen, A.; Khalili, H.; Virkajärvi, P. Concentrate supplementation responses of the pasture-fed dairy cow. *Livest. Sci.* **2006**, *104*, 292–302. [CrossRef]
43. Gouvêa, F.L.R.; Cardozo, L.L.; Canal, J.; Troncarelli, M.Z.; Pantoja, J.C.F. A descriptive study of teat morphology, milking machine characteristics, and milking practices in a sample of Brazilian dairy herds. *Livest. Sci.* **2020**, *241*, 104196. [CrossRef]
44. Candiotto, L.; Adelaide, D.; Elejalde, G. Milk quality in small farms from Southern Region of Brazil. *Cienc. Rural* **2020**, *50*, 1–5. [CrossRef]
45. Giordano, J.O.; Kalantari, A.S.; Fricke, P.M.; Wiltbank, M.C.; Cabrera, V.E. A daily herd Markov-chain model to study the reproductive and economic impact of reproductive programs combining timed artificial insemination and estrus detection. *J. Dairy Sci.* **2012**, *95*, 5442–5460. [CrossRef] [PubMed]

46. Taylor, P.D.; Fahrig, L.; Henein, K.; Merriam, G. Connectivity is a vital element of landscape structure. *Oikos* **1993**, *68*, 571–573. [CrossRef]
47. Tischendorf, L.; Fahrig, L. On the usage and measurement of landscape connectivity. *Oikos* **2000**, *90*, 7–19. [CrossRef]
48. Pascual-Hortal, L.; Saura, S. Comparison and development of new graph-based landscape connectivity indices: Towards the priorization of habitat patches and corridors for conservation. *Landsc. Ecol.* **2006**, *21*, 959–967. [CrossRef]
49. Saura, S.; Torné, J. Conefor Sensinode 2.2: A software package for quantifying the importance of habitat patches for landscape connectivity. *Environ. Model. Softw.* **2009**, *24*, 135–139. [CrossRef]
50. R Core Team. *R: A Language and Environment for Statistical Computing*; R Foundation for Statistical Computing: Vienna, Austria, 2019.
51. Fox, J.; Weisberg, S. *An R Companion to Applied Regression*, 3rd ed.; SAGE: Newbury Park, CA, USA, 2019; ISBN 978-1544336473.
52. Zuur, A.F.; Ieno, E.N.; Elphick, C.S. A protocol for data exploration to avoid common statistical problems. *Methods Ecol. Evol.* **2010**, *1*, 3–14. [CrossRef]
53. Fox, J.; Monette, G. Generalized collinearity diagnostics. *J. Am. Stat. Assoc.* **1992**, *87*, 178–183. [CrossRef]
54. Soares-Filho, B.; Rajão, R.; Macedo, M.; Carneiro, A.; Costa, W.; Coe, M.; Rodrigues, H.; Alencar, A. Cracking Brazil's Forest Code. *Science* **2014**, *344*, 363–364. [CrossRef]
55. dos Santos, G.; Bittar, C.M.M. A survey of dairy calf management practices in some producing regions in Brazil. *Rev. Bras. Zootec.* **2015**, *44*, 361–370. [CrossRef]
56. Ranieri, V.E.L. Reservas Legais: Critérios para localização e aspectos de gestão. Ph.D. Thesis, Universidade de São Paulo, São Paulo, Brazil, 2004.
57. Kleijn, D.; Bommarco, R.; Fijen, T.P.M.; Garibaldi, L.A.; Potts, S.G.; van der Putten, W.H. Ecological Intensification: Bridging the Gap between Science and Practice. *Trends Ecol. Evol.* **2019**, *34*, 154–166. [CrossRef]
58. Weibull, A.-C.; Ostman, O.; Granqvist, A. Species richness in agroecosystems: The effect of landscape, habitat and farm management. *Biodivers. Conserv.* **2003**, *12*, 1335–1355. [CrossRef]
59. Popescu, V.D.; Hunter, M.L. Clear-cutting affects habitat connectivity for a forest amphibian by decreasing permeability to juvenile movements. *Ecol. Appl.* **2011**, *21*, 1283–1295. [CrossRef] [PubMed]
60. Concepción, E.D.; Díaz, M.; Kleijn, D.; Báldi, A.; Batáry, P.; Clough, Y.; Gabriel, D.; Herzog, F.; Holzschuh, A.; Knop, E.; et al. Interactive effects of landscape context constrain the effectiveness of local agri-environmental management. *J. Appl. Ecol.* **2012**, *49*, 695–705. [CrossRef]
61. Donald, P.F.; Evans, A.D. Habitat connectivity and matrix restoration: The wider implications of agri-environment schemes. *J. Appl. Ecol.* **2006**, *43*, 209–218. [CrossRef]
62. Lima, F.P.; Bastos, R.P. Perceiving the invisible: Formal education affects the perception of ecosystem services provided by native areas. *Ecosyst. Serv.* **2019**, *40*, 101029. [CrossRef]
63. Gerosa, S.; Skoet, J. Milk availability Trends in production and demand and medium-term outlook. *ESA Work. Pap.* **2012**, 1–40. [CrossRef]
64. Batista, E.; Soares-Filho, B.; Barbosa, F.; Merry, F.; Davis, J.; Van Der Hoff, R.; Rajão, R.G. Large-scale pasture restoration may not be the best option to reduce greenhouse gas emissions in Brazil. *Environ. Res. Lett.* **2019**, *14*. [CrossRef]
65. de Léis, C.M.; Cherubini, E.; Ruviaro, C.F.; Prudêncio da Silva, V.; do Nascimento Lampert, V.; Spies, A.; Soares, S.R. Carbon footprint of milk production in Brazil: A comparative case study. *Int. J. Life Cycle Assess.* **2015**, *20*, 46–60. [CrossRef]
66. Lambin, E.F.; Meyfroidt, P. Global land use change, economic globalization, and the looming land scarcity. *Proc. Natl. Acad. Sci. USA* **2011**, *108*, 3465–3472. [CrossRef]
67. Santos, J.S.; Feltran-Barbieri, R.; Fonte, E.S.; Balmford, A.; Maioli, V.; Latawiec, A.; Strassburg, B.B.N.; Phalan, B.T. Characterising the spatial distribution of opportunities and constraints for land sparing in Brazil. *Sci. Rep.* **2020**, *10*, 1–11. [CrossRef]

MDPI
St. Alban-Anlage 66
4052 Basel
Switzerland
Tel. +41 61 683 77 34
Fax +41 61 302 89 18
www.mdpi.com

Sustainability Editorial Office
E-mail: sustainability@mdpi.com
www.mdpi.com/journal/sustainability

www.ingramcontent.com/pod-product-compliance
Lightning Source LLC
Chambersburg PA
CBHW061622210926
51850CB00009B/2408
9783036580821